The Species Maker

The Species Maker

{ a novel }

Kristin Johnson

THE UNIVERSITY OF ALABAMA PRESS

Tuscaloosa

The University of Alabama Press
Tuscaloosa, Alabama 35487-0380
uapress.ua.edu

Inquiries about reproducing material from this work should be
addressed to the University of Alabama Press.

Typeface: Adobe Caslon Pro

Cover design: Lori Lynch

Cataloging-in-Publication data is available from the Library of Congress.
ISBN: 978-0-8173-6015-3
E-ISBN: 978-0-8173-9371-7

For my students

Contents

Part IV. *The Historians*

Part V. *Trials and Tribulations*

Author's Note

In 1921 three-time Democratic presidential candidate and former secretary of state William Jennings Bryan began a campaign to get evolution out of American public schools. That campaign resulted in the passage of various state laws that prohibited the teaching of evolution. In 1925 the legislature of the state of Tennessee, for example, passed the Butler Act, which declared it unlawful to teach "any theory that denies the story of the Divine Creation of man as taught in the Bible, and to teach instead that man has descended from a lower order of animals" in any of the state's public schools, including universities. Supported by the American Civil Liberties Union, a high school teacher named John T. Scopes agreed to test the constitutionality of the Butler Act, and so, in the summer of 1925, he was the defendant in what would become known as the Scopes Trial. Bryan served on the prosecution and Clarence Darrow, a well-known agnostic, served on the defense. The trial was broadcast live across the nation, a first for American radio.

The Scopes Trial often comes up in my courses on the history of science. Most students come to class with this event (and the trial of Galileo) in their minds as clear examples of an inevitable conflict between science and religion. But the past, or the present for that matter, can rarely be captured in such stark, simplistic terms. The Scopes Trial was in fact a complicated, dramatic proxy for conflict between several competing visions of the proper relationship between science, politics, and faith. I wrote *The Species Maker* to provide an alternative means of experiencing this complexity. It tells the story of a fictional taxonomist (a scientist who names, describes, and classifies animals and plants) navigating the shifting landscape of American science and society during the years prior to and including the trial. This book is not a detailed history of the Scopes Trial (the trial shows up only via radio, and only in the final four chapters), but rather an imaginary journey through ideas, trends, questions, and debates that culminated in that famous event. The main characters (Martin, Josiah, Ben, Helen, Will, Phoebe, Pete, and Rebecca) are fictional, yet their hopes and struggles are based in tensions that existed within science and American life by the 1920s. Information on the historical figures appearing in the text, as well as the scholarly sources on which stances, dialogue, themes, and events are based, is provided in the notes section at the end of the book (an asterisk in the text signals the presence of a note). For additional supplementary material, including a character map, glossary, historical background, and discussion questions, visit www.thespeciesmaker.com.

My hope is that the book will be of interest to anyone—whether student, scholar, or general reader—interested in the relationship between science and society in the United States, both in the past and in the present. Although set almost one hundred years ago, *The Species Maker* wrestles with many issues that continue to confront scientists and the public in the present day. For example, why are many Americans skeptical of scientists and scientific claims? What are the best means of obtaining true knowledge about our past and ourselves? How can we tell when we are reading our biases into nature? What are the implications of evolution for determining our values and finding both meaning and purpose in life? On what grounds should behavior be governed and decisions be made? Should our understanding of evolution and genetics influence human decision making? What are the proper boundaries between science, politics, and faith, and what are the respective roles of scientists, ethicists, theologians, politicians, and so on? How should scientists balance the duties of activism with their methodological ideals of caution and skepticism? Answering these questions was (and still is) complicated by a crucial, additional question: Who gets to say and on what grounds? Amid the vast complexity of this past, one thing can be said with certainty: none of these questions, or the debates associated with trying to answer them, are going away anytime soon.

I am deeply grateful for the support and suggestions of Tiffany Aldrich MacBain, Greta Austin, Keith Bengtsson, Michael Crow, Shannon Dixon, Erik Ellis, Tamra Erickson, Claire Lewis Evans, James Evans, Paul Farber, Kelly Finefrock, Amy Fisher, Maya Friedman, Alexander and Shelly Garzon, Melinda Gormley, Mott Greene, Shauna Hansen, Donna Johnson, Kim Kleinman, Laura Krughoff, Jane Maienschein, Elizabeth Nielsen, Bob Nye, Mary Jo Nye, Lynn Nyhart, Justin Roberts, Leslie Saucedo, Melanie Siacotos, Katherine Smith, Max Smith, Addie Taylor, Justin Tiehen, Alison Tracy Hale, Ariela Tubert, Lia Van Streeter, Peter Wimberger, my students in STS 370 (Science & Religion: Historical Perspectives) and STS 330 (Evolution and Society since Darwin), and two anonymous reviewers.

The Species Maker

The first step in wisdom is to know the things themselves. This notion consists in having a true idea of the objects; objects are distinguished and known by classifying them methodically and giving them appropriate names. Therefore, classification and name-giving will be the foundation of our science.

—CARL LINNAEUS, *Systema naturae* (1735)

Part I

The Naturalists

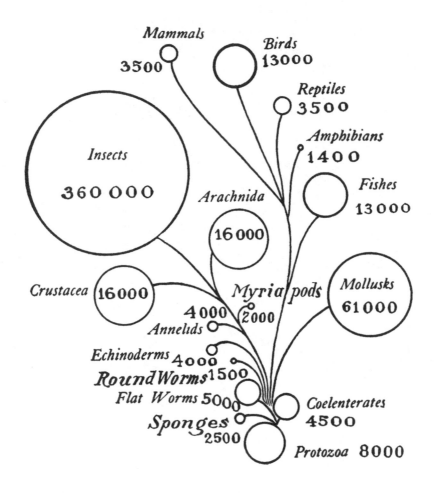

"The Evolutionary Tree" diagram from George William Hunter's *A Civic Biology* (1914).

Chapter 1

It was a marvel of the Age of Science. Thanks to the newswires of the Associated Press, Americans across the country could read newspaper articles the same day they appeared back east. On July 29, 1925, for example, one might learn about disarmament efforts in Europe. Or that automobiles would soon outnumber horses on American roads. Or read the final speech of the Great Commoner, former secretary of state William Jennings Bryan, the same day it appeared in the *New York Times*. A brief background headed the speech:

> The anti-evolution speech, in the delivery of which William Jennings Bryan hoped to make his 'supreme effort,' today was given to the world, despite the fact that its author's lips had been sealed by death. . . . The address was to have been delivered in the trial of John Thomas Scopes, convicted of violating Tennessee's law prohibiting the teaching of evolution in its schools, but by agreement between counsel closing arguments were dispensed with.*

Across the nation many newspaper readers no doubt nodded in sympathy with Bryan's posthumous warnings that morality would collapse if the country continued to allow evolutionists' atheistic "guesses" to be taught in American schools. Others scoffed and dismissed the old politician's speech as the unenlightened result of his ignorant, medieval version of Christianity.

In a small drugstore in the village of Friday Harbor, on an island in the middle of Washington State's Puget Sound, one man stood very still, with neither nod nor scoff, as he read Bryan's words. Professor Martin Sullivan's unkempt brown hair and lack of a jacket were good signals that he was one of the University of Washington biologists (the locals called them Bugs) who descended on San Juan Island each summer. His colleagues at the Puget Sound Biological Station, a mile outside of town, had sent him to retrieve the newspaper so they could dissect Bryan's parting shot. Now Martin stood silently, brow bent in concentration as he read, unaware that the shopkeeper was eyeing him with annoyance because he'd failed to hand over two cents before opening the paper.

Why didn't Martin scoff in disdain at Bryan's warnings? After all, it was his livelihood at stake if Bryan convinced Americans that evolution was too dangerous to teach to their children. By Bryan's measure, Martin and his colleagues destroyed moral responsibility, paralyzed all efforts at reform except the scientific, eugenic breeding of human beings, and justified the might-is-right ethics that had led to the Great War. In blaming biology for all that was wrong with the world, Bryan did not spare Martin's own specialty. He ridiculed those who spent their days in the corridors of museums naming and classifying species by criticizing an evolutionary tree from the biology textbook used by John T. Scopes:

> What shall we say of the intelligence, not to say religion, of those who are so particular to distinguish between fishes and reptiles and birds but put a man with an immortal soul in the same circle with the wolf, the hyena and the skunk? What must be the impression made upon children by such a degradation of man?

Such teachings, Bryan thundered, would drag humanity downward into a struggle of tooth and claw, destroy the unity of humanity, and end any hope of world peace. Yet public high schools and colleges stocked their shelves with such textbooks, allowing teachers to poison the minds and morality of American youth with these materialist doctrines, and all at taxpayers' expense!

Martin knew there wasn't much new in Bryan's speech, despite the newspaper's claim that he had prepared it especially for the trial in Tennessee. The old statesman had been making these arguments since launching his crusade against teaching evolution in public schools a few years after the war ended. Martin tried to remember whether he had scoffed at Bryan's arguments when they first appeared in the papers. He didn't think so. He wasn't in the habit of judging, much less scoffing, quickly. But he did know he had not taken Bryan's crusade very seriously. Passionate debates about God and evolution had always seemed the distracting business of strangers. The dilemmas raised by genetic explanations of human virtues and vices had not touched him. And he had never felt the kind of all-consuming love that, according to Bryan, Darwinian explanations of the heart and mind would destroy. No doubt he had quickly returned to his beetle specimens, diligently paying little attention to what his work might mean for the struggles of his fellow *Homo sapiens* to understand and govern themselves. But now, with this speech in his hands, he felt as though Bryan was standing before him with a posthumous demand: "Retreat safely back to your museum, professor. Or take a stand for everything you love. Show me, if you can, that I am wrong."

—

Two years ago and 2,500 miles away, Martin Sullivan worked long hours at Harvard University's Museum of Comparative Zoology, where the museum's red brick walls conveniently blocked any time-consuming distractions of human troubles and concerns. He spent his days bent over a microscope documenting geographical variation in a large family of weevils, fixing errors in the names and descriptions created by prior naturalists, and sometimes describing and naming previously unknown forms. His ability to ignore human problems and focus on those little animals meant he got a lot of good work done. Which was why his boss at the museum, Samuel Henshaw, was so distraught when, in the summer of 1923,

Martin suddenly quit his job as junior curator of insects to take a position out west at the University of Washington.*

"Good heavens! Seattle?" Sam said, sitting back in his chair. His desk was tucked in a corner of the large room that held the museum's insect collection. Martin stood before that desk as though on trial. "But you've got so much work to do here! What about all those new specimens that just came in from Panama? Must be a dozen new species! New genera even!"

Sam placed a hand against his forehead, and then fiddled with his thin gray hair in a quick, frustrated movement, as though trying to work out the answer to his own question. Martin had never seen his unflappable boss so upset.

"I'm sorry, sir," he replied. "I've already accepted the position."

"And what about your classification of fungus weevils? How in heaven's name are you going to get that done without access to the Smithsonian?" Sam gestured toward the rows and rows of insect cabinets a few feet away from them. "Without all this?"

They both knew the answer, so Martin said nothing in reply. He realized he was holding his breath, hoping Sam wouldn't press him for an explanation of why he was willing to give up access to the best insect collections in the country.

"But why? I know you, Sullivan. You love this work. Everything's got to have an economic angle out west. You'll get roped into killing insects with a bunch of chemicals. You won't be able to do science anymore, I can tell you that much."

Martin knew that if Sam only looked up from his desk once in a while and listened to museum gossip, he'd have come up with some good guesses as to why Martin was leaving. He might, for example, suspect that Martin's father had finally crossed a line, and Martin had to get him out of Cambridge. Permanently. Whether it meant taking a job controlling insect pests or no. Perhaps then Sam would have let him alone.

"This all concerns some girl, doesn't it? Someone's fascinated you with her cooking and she's dragging you west for her health."

"No, sir."

Henshaw stared at a tray of specimens on his desk for a moment and then took a deep breath.

"I'm sorry, Sullivan, but this is a bit of a blow. You're the only one of the younger set who thinks as I do. The others are obsessed with proving themselves useful, and what do they do? Spend a few years studying fruit flies and then go out and spout a lot of nonsense about how biology's going to solve all the troubles of the world."*

"It's a teaching post," Martin said, grasping at anything that might alleviate Sam's distress. "Introductory zoology and general entomology. I'll also be in charge of their insect collection. So I can at least promise to make sure the students can identify more than fruit flies."

Sam shook his head.

"Nowadays students want science to relate to human beings. To their own little problems. Everything's changed since the war, Sullivan. You won't be able to study insects for their own sake. The kids are going to demand you tell them what biology says about human life and destiny."*

Martin smiled. He and his boss had never discussed his preference for studying nonhuman animals, but as a good observer, Sam had diagnosed his weakness.

"I trust myself to be able to lecture for hours on beetles," he replied. "Insects. Even a few vertebrates. But not human beings."

Sam's gaze turned sympathetic.

"Then try and keep up your museum work, Martin. Make sure you can do at least a few hours a day. With no application. No relevance to teaching. No utility. Just sorting things out. And get out into the field when you can. That's the best way to remember how little we know. About most things." A stern frown replaced Sam's sympathetic gaze as he added, "And don't you dare get caught up in this debate over teaching evolution. We've got enough work to do without having to defend evolution against the likes of old Bryan."

Martin gave one more "Yes, sir," and Sam let him go with a final word. "Mind, I expect you to change your mind within the year. You won't be satisfied with the kind of work men do without a good collection at hand."

When he returned to his own desk on the other side of the insect room, Martin leaned back in his chair for a long moment, staring at the bookcase full of entomological journals on the wall opposite. It was official now. He only had a few days left in this beloved, specimen-filled building.

With a deliberate movement, he drew his chair forward, as though the motion could push regret and grief out of his mind, and bent over his microscope. If he worked until midnight, he might be able to finish the description of geographical variation in a species of fungus weevil. Previous taxonomists had named a dozen forms of the specimens in front of him as different species, but after studying hundreds of specimens for weeks, Martin was pretty sure they'd imagined the boundaries. Because they didn't have enough specimens, and hadn't looked closely enough. But with all those specimens and enough time to study them, he knew he might eventually say with certainty, "I know what I have here." This careful description of variation was an almost-sacred act for Martin—the only way to justify having immobilized hundreds of living creatures forever.

To ensure he'd named and classified the species correctly, he had to hunt through old entomological journals and then compare the specimens on his desk with long-dead entomologists' attempts to capture the essence of each little animal with a few lines of Latin. It was an overwhelming task. If, that is, the taxonomist paid attention to all the traps set by his own judgment, and resisted the temptation to decide on a name too early. Martin was four years into studying thousands of specimens of these little fungus weevils (a Swedish naturalist, Gustaf Johan Billberg, had christened the group Anthribidae in 1820), and sometimes felt as if he knew less about the little creatures than when he began.

Most of the creatures beneath his fingertips had been named by men who believed they were the special, independent creation of God. The process of sorting out the names was in many ways still the same: notice as much as possible, document it, discern the differences and similarities that mattered for getting the names right, and place each little animal in its proper place in the Linnaean classification system. For Martin's creatures of choice, most of the categories were easy to discern.

Kingdom: Animalia.

Phylum: Arthropoda.

Class: Insecta.

Order: Coleoptera.

Family: Anthribidae.

Then, at the levels of genus and species, doubt and uncertainty set in. Unable to give names to things he could not define, Martin had collapsed dozens of species into varieties because their boundaries blended into one another once he had enough specimens. He knew that Charles Darwin, who had been trained as a young man to view varieties as formed by the environment and species as created by God, had cited such taxonomic dilemmas as evidence that both varieties and species represented different degrees of divergence from a common ancestor. For why, if varieties and species arose via different processes, was it so difficult to distinguish between them? Thus taxonomists' challenges in naming God's species had become an argument that species were in fact the products of a purely natural process. But whether taxonomists thought they were naming the creations of God or Nature, it was still slow, tedious work, requiring hundreds of specimens and a great deal of time.

The smell of the naphthalene used to preserve all those specimens caused sensitive museum visitors to hold handkerchiefs over their noses. Martin didn't notice the strange perfume anymore. He was very good at ignoring unpleasant things in the interest of getting good work done.

His mother, Mary, had first taught him to attend so carefully to insect life. The memory was one of the clearest of his childhood. She had given him Anna Comstock's *The Handbook of Nature Study* for his tenth birthday and they'd carried the book to a pond.*

"Sit just here, next to me, Marty," she had whispered.

He was entranced by his mother's close attention to the dragonflies flying about them. Then she suggested he write some notes, on what he had seen, heard, or smelled, in a little notebook that she drew out of a pocket in her skirt. She did not ask him, on their walk home, what he had noticed. He, as a result, did not ask her. But she looked through his notebook that evening, and communicated with a smile as she turned the pages that she understood, and had witnessed it all too.

During subsequent visits to that pond, she read Anna's descriptions of the "absurd errors" and prejudices popularly held about insects. She read one passage about dragonflies aloud to him more than once, adding, "Who could be so silly as to believe that they could sew up ears or that they could bring dead snakes to life!" Mary's tone of voice made the fact that humans could accept such ignorant accounts of the creatures around them seem an issue of justice. And what, she would say, did anyone really know about them now?

One evening when he was twelve Martin's parents gave him a microscope, and he first looked at a flea, arrested with some whiskey by his father, under its lens. The revelation of what could be seen beyond the naked eye by immobilizing the creatures on that little plate of glass astonished Martin. Of course, all the little facts he learned about that flea that evening had been observed by others, but *he* had never known them.

In 1910 Martin packed up his microscope to attend Cornell University so he could study with Anna Comstock's husband, the entomologist John Comstock. Professor Comstock disciplined his love for insects into the careful methods of scientific taxonomy. Martin learned how the form and color of specimens held clues to what beetles had been and how they had changed in the past, and how, with enough hours and specimens, he might even be able to record evolutionary relationships with a few Latin names.*

He learned quite early that this work seemed strange to the uninitiated. More than once, his father, Will, stood over his shoulder and asked in a bemused tone, "What's this all good for, Marty? Why'd you choose these little beasts?" His mother would laugh and demand, "Why'd you choose me instead of someone else?" When Will confessed anxiety that pinning all those helpless little creatures would undermine his son's ability to love living things, Mary had replied, "To study something so carefully is a form of love." Amid the ambitious arguments his colleagues gave in defense of their work, Martin preferred Charles Darwin's belief that an instinct for truth, knowledge, and discovery provided reason enough for any scientific research. But he also knew his colleagues' pragmatic reply, had he tried to give such abstract defenses of his days: "Yes, well, Darwin had a private fortune and lived in the nineteenth century. You don't."*

Both facts were hard to forget. For this was the very modern twentieth century after a devastating war, when every institution and discipline was called upon to "stand and deliver" and help put the world back together. The men at the United States Department of Agriculture's Bureau of Entomology defended the study of "these little beasts" on the grounds entomologists could prevent millions of dollars of damage to agriculture and timber, and stop the spread of dangerous, insect-borne diseases. Martin had put in his time working for the bureau. During the war he was in charge of grain inspection at Bush Terminal in Brooklyn, after Comstock recommended him to the director of the bureau, Leland Howard, as a good weevil man. The voracious appetite of a few species threatened the enormous quantities of wheat stocked up for shipment to the Allies in Europe. So Martin, who studied beetles because he was fascinated by their diversity, became a soldier on the front lines of a battle to kill a few kinds. It was dusty, noisy work, and his team of exterminators killed insects on the spot. But then, as Martin often reminded himself during those hours and hours of walking the dock warehouses, killing on the spot and working indoors had constituted scientific entomology for centuries. The main difference was that at the museum the dead specimens were neatly stored away in cabinets, carefully labeled for the sake of future study, rather than tied up in bags and thrown into the garbage.*

He could have stayed at the Bureau of Entomology after the war. Even before the guns in Europe had fallen silent, Howard was insisting that the chemicals and airplanes developed during the war to kill men should be aimed at humanity's insect enemies. The next war would be a "War against the Insects," and the nation needed a scientific army made up of men willing to join the fight. The campaign worked. By 1923, the bureau employed an army of four hundred entomologists at more than eighty field laboratories, with stations that could be mobilized to the site of an insect outbreak in just a few days.*

Martin's decision to leave Howard's army of applied entomologists at the war's end and take a position at a natural history museum at Harvard had baffled his more practical friends. Museums and the taxonomists who stalked their mothballed corridors had been ridiculed for decades as hidebound and old-fashioned, and such criticisms had only heightened in recent years. Since Martin had arrived at Harvard, more than one geneticist had derided the museum in his presence and wondered why administrators allocated so much of the Zoology Department's budget and floor space to millions of dead things. Some of the other museum curators got into heated arguments with colleagues who thought only experimental methods counted as science, and only scientists at lab benches deserved funding. Martin just shrugged and got back to work when these debates broke out. It was one of the things Sam liked about him. "You're right that it's a waste of time to argue with such dogmatism," Sam would say. "Best to just focus on doing good science and prove by robust research that these faddists are wrong to insist there's only one way to do things."*

Martin believed Sam's defense of his reticence gave him more credit than he deserved. Sam didn't know that since college Martin had been unable to debate most issues that animated his friends without developing an embarrassing trembling in his voice and hands. He had once heard a professor of physiology lecture on how stress released a whole cascade of autonomic responses until adrenaline coursed through the organism's veins, preparing it for either fight or flight. The professor pronounced that this mechanical response represented just one important element of the "beast within" that determined human tendencies and behaviors, from adultery to war. Martin sometimes wondered whether his system had misdirected an evolved response to physical danger to intellectual fights. What, he thought, would the effect on his frame have been in a real battle, in a trench in France?*

Martin had worked at his desk for several hours after his interview with Sam Henshaw when he heard the words "Christ, Sullivan" and looked up. Sam stood in front of his desk, hat in hand and coat on.

"What a gift you have to concentrate when the world falls to pieces about you."

"Might just be a curse," Martin replied with a slight smile.

"Comstock's got you all networked, I guess? Don't need any of my help?"

"Yes, sir. I suspect he's how I got the job in the first place. But thank you."

"You said you'll be teaching mostly?"

"I think so."

"Well. No doubt it's better to have someone out there who can teach kids to try and understand the poor beasts rather than how to kill 'em." He put on his hat and asked one more question. "Your dad going with you?"

"Yes sir."

That was pretty good evidence Sam hadn't "asked around" and still had no idea why he was leaving.

———

When Martin first told his dad he'd been offered a position out west, Will had looked up from his newspaper with undisguised excitement.

"New things might really be done out there, Marty!" he said.

Neither of them expected to leave Cambridge. Martin had only mentioned the offer for the sake of conversation. Now, an hour after his interview with Sam, Martin announced as calmly as he could, "I'm going to take the post in Seattle, Dad. If you'll come with me."

Will's hand fumbled in his breast pocket to retrieve his glasses so he could see his son clearly. Martin did not want to look into his dad's searching green eyes at that moment. But he did so in the hopes a gaze might still the questions. So father and son regarded each other across the table, almost a mirror of each other in appearance. Both were of medium height, and any physical differences were slight, the mixture of the passage of time and habit. A high percentage of Will's brown hair had turned gray, and the edges of his eyes had been permanently wrinkled by his easy laugh. Martin's forehead bore the traces of long hours of concentrating at his microscope. He did have the capacity for a broad smile that spread to his eyes, like his father, though Will had not often witnessed that full, open smile in the past few years. The creases on his son's brow worried him. They were why he gave in so quickly.

"All right, Martin. Maybe out there your old man can be respectable and keep out of trouble."

Martin smiled. "That's doubtful." Then he added, "It's okay, Dad."

He turned back to an old monograph on the genus *Ormiscus* and Will pretended to read through a stack of papers. As a low-level attorney who had worked at the same law firm for thirty years, Will spent most evenings bent over the files of clients who couldn't actually pay for his services. When he wasn't home, it meant there was trouble somewhere. He was on call for the American Civil Liberties Union whenever the local police refused a labor organization access to a meeting venue, and he was often the first attorney on the scene when the police chief tried, yet again, to shut down the Birth Control League office for distributing "obscene" literature.

The apartment father and son shared near the museum was packed tight with Will's library of books by freethinkers: Thomas Paine. Karl Marx. Friedrich Engels. Charles Darwin. Eleanor Marx. Robert Ingersoll. Mary Wollstonecraft. John Draper. Andrew Dickson White. Thomas Henry Huxley. Will loved this library, for he believed it tracked the steady advance of science and reason in the face of superstition. He, too, believed knowledge was a matter of justice and that greater knowledge would inevitably lead to greater good. It was one of the things that had united Martin's parents when they met as undergraduates at Cornell in the 1880s.

Martin's favorite childhood memories, apart from those walks to the pond with his mother, were those times when, at their cramped but tidy dinner table, Mary would take on her husband's enthusiastic pronouncements about some book or idea. Will often needled Mary with grand plans for radical social reform, just to hear her eviscerate his arguments with a barrage of well-targeted questions. There had been a strict rule in the Sullivan household that, at dinner parties, Mary must be given as much time as the men in conversation. Any guest who did not obey this

{14} *The Species Maker*

rule was not invited again. Martin learned from an early age that his parents' arguments were in fact just for fun, either because they really agreed at base or because Will fully expected to be bested.

Martin had witnessed only two real fights between them, the first when he was about ten years old. He didn't know until much later that it involved their very different judgments of one of Will's heroes, the atheist socialist Edward Aveling. Will loved to tease Mary, who called herself an agnostic, with Aveling's quip that agnostic was really just atheist "writ respectable." Mary would reply, "Well. Darwin said atheism is merely agnostic writ aggressive. You can pretend all you like, my dear, but you've got too much of the pacifist in you to join your beloved Aveling's army."*

Then Aveling's lover, the brilliant socialist Eleanor Marx, had committed suicide. Martin hadn't understood why silence descended upon the apartment for days. But later he learned his dad, who revered Karl Marx's daughter for her radical commitment to remaking the world along the lines outlined by her father, blamed Aveling for her death. Gossip had revealed that Aveling had married a young actress in secret and broken Eleanor's heart. His dad got rid of all Aveling's books and switched to reading Martin lectures by the Great Agnostic, Robert Ingersoll, instead, which meant he got a much heavier dose of agnosticism than atheism as he grew up.*

Martin could never quite figure out what had divided his parents in the face of this tragic love affair between two people neither of them had ever met. But sometimes, when he was in a rare, hereditarian mood, he wondered if his tendency to shrug and withhold judgment on most things had come from his mother. Perhaps she had been unwilling to indict either party because they really knew so little about the case, and his father had been unable to tolerate or understand her view of the matter.

The second conflict, more a simmering tension than an open argument, happened during the war, when Martin was old enough to understand things better. To call for reform at home in the midst of a war looked suspiciously un-American, and Will was called a Bolshevist in the street more than once. It was a damning accusation at a time when politicians were proclaiming, "My motto for the Reds is S.O.S.—ship or shoot," and insisting that communists should be gathered up and placed in a ship of stone with sails of lead and hell as their first stopping place. Martin and Mary had both feared for Will's safety, and Mary's requests that Will be more careful irritated her husband to no end. That is, until Mary's illness set in and Will was housebound for those terrible months, unable to do anything but sit beside her and wait.*

After Mary's death, Martin did not begrudge his father the various causes and crusades, though he had many an anxious evening when, during the Red Scare that set in after the war, Will failed to come home some nights. By 1923 the policing of radicals and reformers seemed to have died down, but Martin still breathed easier when his dad came home right after work, even though it meant constant interruptions in whatever reading he'd hoped to get done before bed. Keenly aware of the absence in the room since his mother's death, Martin had resolved to endure Will's habit of tossing news articles and labor pamphlets at him.

"You've got to look up from your bugs now and then," Will would say.

Martin tended to smile, read whatever his dad tossed at him, and reply quite briefly. Sometimes he shrugged, though with his father it wasn't fear of adrenaline that inspired the movement, but honest ignorance as to what to think about the latest moves by the League of Nations, the wisdom of prohibition, or the fate of a nation that tolerated flappers.

Will used to lose his temper in the face of his son's shrugs. He took tremendous pride in the fact Martin was a scientist. It meant his son was doing his part to apply reason rather than superstition to his understanding of the world. But he could never understand why Martin had chosen to study a group of obscure beetles, or why he refused to be drawn into debates about human animals. Now that his son was in his midthirties, they'd reached a bit of a truce with respect to each other's ways. Will knew Martin's response when pressed: "I just want to get good work done, Dad. I don't want other men's crusades to determine my own days."

For the past few years, Will's inability to understand his son's decisions tended to manifest primarily in teasing. On their last night in Cambridge, with boxes of books strewn about the apartment floor ready to be shipped west, he began chuckling over a page of the *New York Times*.

"Listen to this, Marty. It's right up your alley," he said and proceeded to read aloud.

The writer W. L. George confessed last night, that he collects cases of women—technical, not wooden cases, that is—and catalogs them, puts them on the end of a pin and examines them as if they were interesting insects. It is much better to be stuck on a pin than to be ignored, said Mr. George, with a suavity that caused some giggles among those present. At present he has sixty-five distinct species of woman safely tucked away.*

Martin did not look up as he replied, "I wager if more taxonomists had taken on that particular specialty, we wouldn't have to work so hard to convince people we're useful."

"Perhaps not. Still. Think of the potential arguments about how to produce a scientific classification and what characteristics to choose! Those would be some fights worth witnessing! And if taxonomists decide to place the different types of females on some hierarchy of progress, with each man's little favorites placed on the top, well then, you might just have to decide on something, my boy."

Martin smiled, but said nothing. When, after some moments, Will spoke again he forced the words out quickly, as though the question had been bottled up for some time.

"Think you'll see Sarah before we go?"

Martin looked up at his father, who was now staring down at the newspaper as though he had just read Martin the weather report.

"No, Dad."

When his dad said nothing in reply, Martin, too, pretended to return to his reading. But he was astonished at this evidence that his dad had apparently been harboring some version of "Think you'll see Sarah?" for so long.

The question inspired vague but painful memories. His mother had died of breast cancer soon after he and Sarah parted. Or was it before? He could never really trust his memory of those days. The final months of crisis had been more dreamlike than real. The demands of just getting through each moment had dissolved days and hours from his mind. And like a dream, he could recall very little. Even how, precisely, his first and only love affair had ended. He had an elusive sense that there was some kind of ultimatum. Perhaps that he decide on certain things. He had never told his dad that Sarah had soon become engaged to someone else.

He remembered just one thing quite clearly from those terrible months, as his parents faced parting from one another: His father's grief-stricken anger, though most days that anger really had no clear target. As an un-believer, Will couldn't blame God. But something seethed within him. Against the universe. It escaped just once, with violence, when one of Mary's well-meaning friends asked a local minister to visit a few days after they lost her. Will took one look at his collar and, before the man said a word, jumped up at him with a beastly roar. He didn't touch the man but he might as well have, the way the poor man fell back into the hallway. Martin had had a hard time restraining his dad and preventing him from following the man out into the street.

"Don't you dare come in here with your damned benevolent gods!" Will shouted before Martin could get him back into their apartment, where he crumpled into a chair like a contrite child.

Martin knew what was happening in his father's mind, or heart, at that moment. His father loved Robert Ingersoll's essay "The Gods," in which Ingersoll ridiculed the old argument that nature demonstrated the wisdom, power, and goodness of God. Oh yes, Ingersoll countered with indignant mockery, cancer is indeed an ingenious adaptation! Then that ingenious adaptation killed Mary, and father and son were powerless to do anything but witness as the tumor spread, was excised, reappeared, and spread once more.*

For days after Mary's death, Will seemed to be in physical pain from the effort it took to force himself to keep breathing. Martin's moments were consumed by taking his dad on long walks and listening to him describe every memory, as though he was carefully organizing them in his brain so he could remember each and every one. Sometimes Martin felt as if they both knew they were trying to keep his dad from going mad.

At some point, Sarah simply disappeared from his memory of those days. She was in his life. Then she was not. It was strange. He could remember the placement of certain insect specimens amid dozens of cabinets and hundreds of drawers. But he couldn't remember how, why, or when the love affair had ended.

Which was why he frowned when his dad, from the safe distance of at least three years and a few moments more of silence, now spoke again and asked why he had let Sarah go. Martin was about to shrug when his dad added, "Was it because of me, Marty? You think you had to take care of me?"

Martin realized for the first time that his dad blamed himself for Sarah's absence.

"For heaven's sake, no, Dad," he said quickly. He tried to alleviate the doubt in his dad's eyes. "Sometimes I wonder whether seeing you and Mom made me unsure of the strength of my own feelings. Sarah was a good observer. She would have seen it."

"Well, that's a high standard, my boy," Will replied as the doubt was replaced with tears. "But it's impossible to reach the stars without jumping first. Usually one must jump quite blindly, and trust that one of those stars is a lucky one."

Martin looked down at the copy of *Transactions of the American Entomological Society* on the table before him. The page was open to a long, detailed description of a new species of Anthribidae from Texas. He knew very well that plenty of individuals could speak from a firm faith that they knew the truth about something more important than the shape and color of a weevil's elytra. Who could even act with confidence and believe they had something worth fighting for. His own reluctance to pronounce "firm conclusions reached" made him an excellent scientist. He knew that. But he sometimes doubted whether it made him a very good human being, who must ultimately commit to something in the face of overwhelming uncertainty.

Chapter 2

On his first day on the University of Washington campus Martin reported to the Zoology Department's offices on the second floor of the Science Building, an imposing, red-bricked structure on the north edge of campus. The director of the department, Trevor Kincaid, met him with a firm handshake and then looked about his office with apparent confusion as to what to do next.

"Hell stirred up with a stick, isn't it?" he said, before exclaiming, "Ah! Here it is!" and moving a giant stack of papers off what then became apparent was a chair. He winked as Martin sat down. "Only God and I know where to find everything in here."*

Trevor was a bit shorter than Martin, with thick, short-cropped graying hair. His kind eyes were close together, framed by dark, curved eyebrows that Martin would soon learn had a tendency to shoot up with curiosity when anyone posed a particularly interesting scientific puzzle.

"I suspect Sam gave you some trouble about accepting a teaching post?" he said.

"Yes, sir," Martin replied. "But we patched things up well enough before I left. I promised to send him some specimens."

"Well, in that case we'll need to get you out into the field. When the Experiment Station requests your aid, oblige them when you can. I'll be sending your name down as an expert on weevils, if you've no objection. It's in a place called Puyallup. You'll get used to the Indian names. A bit of a trek but use the excuse to get out of the city for a few days if you can."

Martin couldn't help but smile. "This is why Sam gave me trouble. Experiment stations, that is."

"Ah, don't you worry now. We've got a field station up north where you can collect to your heart's content, for no reason apart from your own curiosity. But here on campus I've found that focusing on the beasts that are damaging or beneficial is one way to prioritize. Amid the millions we know nothing about. You will see it guides my own time. Lately I've been trying to get a bunch of Japanese oysters to spawn. Our native oyster's been done in by overharvesting. Good to show the men in Olympia we're worth paying for, when we can. But apart from passing your name along I won't impose that demand on you if I can avoid it. Besides, you're going to have enough work to do on the collection."

Trevor took him to the chaotic room that contained the department's insect collection, where a corner had been designated as Martin's office. Cabinets lined the walls, but some of the doors were barely hanging on their hinges, and boxes and specimen trays were strewn about the floor. Having cleared a space for Martin to stand, Trevor stood and surveyed the mess.

"That cabinet contains sixty thousand specimens from the Harriman Expedition. Almost twenty-five years ago, and I've barely had time to touch them. That's after I sent about a hundred thousand to specialists in the east. Most are still unnamed, of course."

Trevor apologized repeatedly for the state of that room, but Martin saw that good microscopic work could be done beneath the large, south-facing window. After Trevor left him alone to get his bearings, Martin sat for hours in that crowded room, in the warm afternoon light, looking over the specimen lists. The locality names were mesmerizing: Snoqualmie. Snohomish. Quilcene. Skykomish. Quillayute. Quinault. Cathlamet. Sequim. Chehalis. Issaquah. Mukilteo. All mixed in with scores of names of European origin, from San Juan to Marysville. Each represented a tiny creature encased somewhere in the boxes strewn about him. There was indeed plenty of work to do here. If he could get everything organized.

Martin soon met the other member of the department: a tall, thin man named Ben Cardiff with thick hair that needed a trim and round glasses pressed high on his nose. He had the disheveled look of a man who knew he was going to be spending his day in a lab and need not impress anyone.

"So you're our new species maker?" he said when Trevor introduced them, but his grip as he shook Martin's hands was firm, and the question was accompanied by a wink.*

So Martin smiled. "From a certain point of view. Sure."

"Harvard man?"

"Cornell." Martin replied. "But I just came from the Museum of Comparative Zoology."

"Jumped ship while you could, eh?"

Ben laughed in response to Martin's confused look. "I'm the department physiologist. Which means it's part of my job to put all you Tall Ship men on notice that the Age of Steam Engines has begun."

Just one week later, fall term began and Martin had to walk into a classroom and introduce himself as the professor, by definition the most knowledgeable person in the room, the one of whom all questions would be asked, and from whom all answers were supposed to be given. He spent the entire hour wishing himself safely back at the museum in Cambridge, where he must pronounce to no one and on nothing. The experience kept him at his desk, prepping for his lectures late into the evenings.

Except Fridays. At the end of each week Ben and Trevor insisted he join them at the Faculty Club at five o'clock. Trevor had warned him that some of the professors would try to determine his political views to see "what he would be good for." He explained the origins of this little game in the 1919 Seattle General Strike, when thirty-five thousand of the city's shipyard workers walked off the job demanding higher wages and a shorter workday and the city had become divided between "Reds" and "capitalists." The capitalists suspected university faculty of being Reds, and the communists suspected professors of being the lackeys of lumbermen and industrialists. There had in fact, Trevor added, been few pure versions of either, but that fact hadn't much mattered in the heat of the Red Scare.*

"In any case," Trevor explained, "the legislature put us on a starvation budget for a year. We got off a bit better than the political science faculty. Seattle businessmen tried to abolish that entire department altogether."

"Well can you blame 'em?" Ben said. "Should have taken history with 'em. Parrington says the sole business of history is to arouse an intelligent discontent and fruitful radicalism."*

"Certainly aren't words to comfort our lumber barons and industrialists." Trevor smiled. "He was fired from the University of Oklahoma by administrators intent on rooting out immoral influences. Calls himself a diluted Marxian."

"Poor Sullivan's going to start thinking the rumors we're all Red out here are true," Ben said with a grimace.

Martin came to like Ben well enough when they were working on some detailed empirical problem, but if anyone on campus could send him to the refuge of the collection room and his microscope during his first year's teaching, it was this passionate, argumentative physiologist.

During his first months in Seattle, Ben often knocked on the door frame with a "How many species made today, Sullivan?"

Martin was used to such jibes. He knew plenty of scientists who openly pitied any well-liked colleague who spent his days sorting specimens of "bugs" in a drawer. It was no use pointing out that he worked on beetles, not bugs. Clarifying the organisms concerned would not parry their primary criticism: that dreaded term "sorting." During their first few visits to the Faculty Club, Ben had tried to get Martin to defend taxonomic work. But Martin soon determined that Ben didn't really care a whit about whether taxonomists survived on university campuses or not. He just liked a good fight.

Martin learned that a good poker face was the best way to keep off of Ben's radar when arguments broke out. During his third visit to the Faculty Club, just as he'd found a moment in which to excuse himself and get back to preparing Monday's lecture on geographical variation, a young man with a moustache and serious gaze came out onto the veranda. The man hesitated visibly when glancing toward the trio of zoologists, but then walked over.

"This is Leslie Spier, from the Anthropology Department," Trevor said by way of introduction as Martin stood to shake hands. "Did his graduate work at the museum in New York. I'm sure you two have some common acquaintances."

But Professor Spier did not seem inclined to indulge in memories about mutual acquaintances. He turned to Ben.

"What's this I hear about you writing an editorial against the campaign to stop military drilling on campus?"

"I don't mind a man opposing military drills," Ben replied. "But I do take issue with a man's reasoning if it's nonsense. The appeal insists that the drilling cease on the grounds training young men in the science of taking human life is subversive of every human instinct. You'd be on firmer ground if you didn't base your arguments on fairy tales about the natural pacifism of man."

Spier shook his head. "I'd be more careful, if I were you, Cardiff. If Bryan is the only one acknowledging that man has extraordinary powers for cooperation and good will, he's going to have a damned sight more men and women on his side."

He went back inside. Ben grunted and mumbled something about how he needed another drink. That was Martin's first clue that there was more than soda water stocked at the club. Ben let a few moments pass, no doubt so Spier was long gone, and went inside as well.

"The lectures on ethnology used to be in our department," Trevor offered in explanation of the little row. "Ben thinks we shouldn't have given them up. Especially not to a man trained by Franz Boas."

That was enough for Martin to determine the point at issue pretty quickly. Martin had never met Professor Boas, but during breaks studying the Anthribidae collection at the American Museum of Natural History in New York he had wandered through Boas's Hall of Northwest Coast Indians. Boas had collected most of the hall's artifacts, including an enormous Haida canoe, carved from a single cedar tree and painted with black and red designs. He'd heard of the fierce arguments between Boas and the director of the museum, Henry Fairfield Osborn, over how Indian artifacts should be displayed for the public. Osborn insisted on displaying a ladder of civilization, while Boas was intent on crushing the idea of ladders altogether.

For years Will had followed Boas's campaign against those who claimed that heredity, rather than environment, determined all. Madison Grant, an influential trustee of the museum, had written a book called *The Passing of the Great Race* in which he derided New York as the sewer of Europe on account of the Jews, Poles, Irish, Italians, and others of "inferior" genetic quality entering in a constant stream at Ellis Island. The laws authorizing the sterilization of criminals and the insane, he declared, must be extended to weaklings, defectives, and "worthless racial types." Boas had written a scathing review of the book, one that Will had read aloud at dinner at least three times since.

Martin had a hard time taking Grant as seriously as his father and Boas did. The man divided *Homo sapiens* into several different subspecies and insisted children of mixed marriages were of a lower racial type. Martin thought the bad taxonomy and shrill tone would surely damn the book. Then the biologist Frederick Woods called Grant's work an "interesting and pioneering attempt to interpret history in terms of race" in the pages of *Science*. His father had tossed that issue at him with a "What are your damned colleagues about, Marty?" Meanwhile, at the museum, Grant had dismissed Boas as a Jew, and then, rumor had it, tried to get him fired. All within the walls of an institution billed as teaching the undisputed facts of science to the American public.*

When Ben returned he was still stewing.

"I don't understand how we could give up the ethnology course to that man," he said, stirring his drink vigorously.

"Come now, Ben. If their theories don't work, they'll fall to the wayside soon enough. Survival of the fittest and all that."

"Their criticisms of Indian policy are going to lead to a good sight more misery than whatever they accuse us of."

Trevor looked at Martin in an attempt to explain Ben's outburst.

"Some of the anthropologists are arguing our Indian policies need to be completely reformed. That, in fact, the Indians might have something to teach us rather than the other way around. They want to close the Indian schools."

"They're spouting a lot of sentimental nonsense about the equal value of all cultures," Ben added.

"I think they're just arguing that each man has the right to determine his own destiny. Even an Indian."

"Fine. I don't begrudge any man, Indian or otherwise, the right to live or think as he likes," Ben replied. "And I won't deny our dealings with them have been dishonorable. When the forest reserves were established any Indian who dared point a gun within the boundaries, of which he had never heard, was taken into custody. We haven't dealt with them fairly. But that doesn't remove the fact that Indians' ways and mores destined them to extinction when Europeans first arrived on these shores. Boasian anthropologists can wish otherwise but wishes won't affect Nature a whit."*

Trevor frowned. "So you agree they should close the Indian schools? Let the weak go to the wall, and all that?"

"No. At this point it's kinder to keep the Indian schools going now we started. Won't help the Indians in the long run, but it's more humane than switching policy midgeneration and turning all the kids out. But it was all a foolish waste of taxpayers' dollars from the beginning. Biology says clear enough it's impossible to reform a race who Nature dictates must fail in the struggle for existence. As though plastering a bit of civilization over years of savagery'll prevent them from dying out."

He glanced at Martin as he spoke. "What's that look for?"

Martin shrugged. "I suspect every man's veneer of civilization is pretty thin at times, whether he's an Indian or not."

Trevor chuckled. "You apply that to yourself, Sullivan?"

But Ben had heard something in Martin's voice that he didn't like.

"Look. I don't like Nature's verdict any more than the next man with an ounce of human feeling," he said. "But to think we can change natural law is to be willfully blind in the face of everything biology tells us. The clear message of evolution is that the only system that secures progress is a system of struggle and self-reliance. That degeneration occurs whenever the range of competition is narrowed or incentive to activity lessened. Any paternalistic policy that interferes with that struggle will increase suffering in the long run."*

Martin just smiled. "No chance the struggle was laid out on unnatural grounds in the first place?"

"Christ, Sullivan, you sound like a communist," Ben replied.

—

Will had a good laugh when Martin repeated Ben's comment. They were sitting at the small table in their new apartment eating some sandwiches Martin had picked up from the corner grocer on his walk home. Trevor had helped them find the apartment: two bedrooms, a kitchenette, and a water closet on the second floor of a house owned by a widow named Mrs. Macleod. Their landlady was a funny mix of Victorian and modern styles. She dressed in dark gray dresses that reached to her ankles, but her hair was cut short and marcelled around her ears and a mess of art nouveau bracelets jangled around her wrists. The house was a similar medley of old and new. The downstairs parlor she offered for their visitors was a confused jumble of old knickknacks, overstuffed cushions, and the latest modern furniture. She was very proud of a new radio in the corner of the parlor, and of a modern, newly installed kitchenette in the upstairs apartment. But it was the view of Lake Washington and Mount Rainier, from a small deck adjoining the kitchenette, that convinced Martin to take the place.

"Heavens, Martin, how did you get on someone's radar as a Red before me?" Will teased.

"I suspect he's pretty certain no biologist worth his salt could actually fight on the side of communists. Even an old-fashioned taxonomist."

"Well. Sounds like you've got a few enlightened colleagues. A man trained by Boas. Now that's something!"

Then he shook his head, chuckling. He took the final bite of his sandwich and stared at the ceiling. After swallowing he looked at Martin with a good dose of affection in his gaze.

"A man can always find a fight if he wants one, Martin. Even, I'm pretty sure, a taxonomist. Of course a fine battle will probably find him if he doesn't choose one first. Even in the most isolated monastery. Could try and be a hermit a hundred miles from other men, I suppose. But a hermit probably just ends up fighting himself. And I wager that can be a damned sight worse." He paused for a moment. "I suspect you won't always have as much control over your moments out here as you like. Trick is to choose by whom you'll be influenced. Other crusaders tend to leave a man alone, once he's associated with at least one battle."

Martin smiled as he reached across the table to clear his father's plate. "That a promise?"

For weeks after their arrival in Seattle, Martin had worried that his dad would be unhappy, torn from all the causes he had loved in Cambridge. But Will had fallen in love with the city soon enough. There was so much that needed fixing, he announced, and so much scope for reform. He'd learned of the Centralia and Everett Massacres, when confrontations between the radical Industrial Workers of the World—the Wobblies— and the powers that be had turned fatal after the war. Of Anna Louise Strong, recalled from the Seattle School Board for her communist views and now traveling the Soviet Union. Of her brother-in-law, Charles Niederhauser, fired from West Seattle High School as a subversive atheist and anarchist because he had tried to get students to understand the Wobblies' perspective. Will regaled Martin with stories of the city's annual Leap Year Feminist Ball, which took as its motto "Woman is Man's Equal in All Things," and of the Seattle Labor College's evening lectures for the working class. Of local bootleggers unloading their cargo onto Seattle's docks in full daylight, and of their use of parliamentary procedure to set prices. And of the "good" bootlegger Roy Olmstead walking the streets of Seattle with a big smile, hailing the city's prominent fundamentalist minister, Rev. Mark Matthews, in the street with the words "Don't take life too seriously, reverend!"*

At one point Will considered taking up the case of a teacher fired for teaching a book titled *Mind in the Making: The Relation of Social Intelligence to Social Reform.* The Seattle School Board had balked at the author's premise that social reform must be based on the fact biology had altered notions of human nature and therefore ethics must be based on a new, scientific basis. But Will decided the young man needed a lawyer with a stronger local network if he was going to win his appeal.*

As he cast about for a cause, he began collecting books on the politics and history of Washington State. Issues of *Sunset*, "the West's Great National Magazine," piled up in one corner. Only editions of the radical *Seattle Union Record* and a small stack of bright red pamphlets in another corner, mysteriously higher each week, made Martin nervous.

Unwilling to give his time to any of the big Seattle law firms because he "knew what kind of work that would be, and damn me if I'll do it," Will co-opted Martin's course catalog and began filling his empty hours with lectures at the university. He clearly developed an intellectual crush on one of the most radical professors, Theresa McMahon in the Department of Economics, after sitting in on just a few sessions of her class on American labor problems.

"She's crazy enough to argue that women have a right to work in whatever field they please," he explained to Martin with mock disdain. "Says they weren't all born cooks and housekeepers, any more than men were all born bricklayers! Imagine that!" His voice fell to a conspiratorial whisper. "I suspect she's one of those feminists causing the downfall of civilization!"

He told Martin of how McMahon had helped get a minimum wage law passed, apparently a weighty enough crime in the eyes of the powerful. But her worst crime was to insist employers had no right to fire a girl for "moral failings."

"The ministers went quite apoplectic when they heard that!" Will added with glee. "She destroyed the university's appropriation a few years ago by testifying in the legislature about the horrendous working conditions in the salmon canneries. Gave a stirring speech about how working in salt water for hours on end destroys the fingers of the Indian women and children. Some of your colleagues warned her that testifying in Olympia's an undignified thing for a professor to do. Undignified or not I bet their real concerns were next year's appropriations. Pretty convenient a-professor-doesn't-do-such-things talk, I'd say."

This was just the kind of woman his dad could turn into a heroine. That worried Martin for a time, until he discerned that Will was just as much in love with Theresa's husband, history professor Ed McMahon, whose students apparently couldn't get jobs teaching history in local schools on account they had been "radicalized" by his irreverent interpretations of the nation's past.

But Will didn't find work via the McMahons' various campaigns. He found it through the aid of Trevor's wife, Louise. Soft-spoken and gray-eyed, Louise was twenty years Trevor's junior. She had been his graduate student and was an excellent naturalist in her own right. When, upon their first meeting, Martin had confessed a strong case of nerves at having to teach the idealistic students he was encountering, Louise had smiled and said, "Then you will do just fine." They had become good friends since then, and Louise insisted he bring his dad to Christmas dinner.

The only other guest was Professor Spier's wife, Erna Gunther. Spier was back east at a conference. Erna was also a student of Franz Boas and had been hired to teach some anthropology courses. But Martin hadn't met Erna at the Faculty Club. Women weren't allowed through its doors.*

She was short, with curly black hair cut into an unruly bob, and had a serious cast to her mannerisms except when playing with her two-year-old son Bobby and the Kincaids' two young daughters, Marjorie and Barbara. Will was delighted to meet someone who had come from some of the biggest fights in academia back east. He peppered her with questions about what Boas was like and then asked about her own work. It was just like his dad, Martin thought, to pay most attention to one of the women in the room. Will's excuse, if teased, was that fear that an old man might be accused of flirting was no excuse for being unjust to a fellow human being.

Though Erna refused to share the stories she was gathering on the Tulalip Indian Reservation north of Seattle on the grounds it was best to hear the stories from the storytellers themselves, she was willing to talk about Seattleites' attitudes toward her work.

"Most people think of the Indians as refugees doomed by some grand civilizing process," she explained. "If they choose to notice them at all. They think I'm collecting stories because the storytellers are going extinct. It's hard to get people to see the tribes as living communities. Or to imagine we might learn something from them, rather than the other way around."*

They had a lot to talk about once they understood each other. For weeks Will had been reading a series of articles in *Sunset* magazine attacking the Bureau of Indian Affairs. The entire government policy, Will lamented, assumed that Indians were at a lower level on the ladder of evolutionary progress than whites. Martin had tried to explain to his father more than once that good taxonomists didn't talk of "higher" or "lower" organisms, whether butterflies or human beings. But Erna made short work of that claim when Will asked her how to fight the bureau.

"Well, for starters, get biologists to stand up against all this nonsense about ladders of civilization," she replied.

"Amen!" Will exclaimed with a laugh, glancing at his son.

"Maybe men had an excuse in the nineteenth century. I don't know," Erna continued. "But we're trying to get the public to abandon the old ladders, and we won't be able to do it without your aid. Not when biologists are giving the ladders scientific credentials and talking about the dangers of lower and higher races mixing. No one's going to pay any attention to what we're saying."*

Will leaned forward across the table. "But don't you think what really needs to be undermined is the assumption that a man is civilized only when he values private property?"*

Erna nodded, pleased Will understood.

"Yes. But we're branded Soviet Indomaniacs when we try. The old missionaries tried to gather all the tribes in a few places and turn them into Europeans, with plots of fenced land and a little bit of savings. Because that seemed like the godly thing to do. Then it became the scientific thing to do. But as soon as the Indians amassed any wealth, they threw a potlatch and gave it all away. Every last article. Most whites take that as evidence Indians are doomed by Nature, rather than that a man's generosity toward his fellows might be the fundamental, organizing principle of a society. I've heard biologists claim there must be some failure in Indians' biological mechanism, because they haven't evolved far enough to value private property and competition. The worst of it is, many of the younger Indians have come to believe the tales. Hard not to when the bureau insists biology backs them up."*

"Ben doesn't speak for all of us." Trevor spoke finally. It was Martin's first clue that the scientists in the Zoology Department did not see eye to eye regarding evolution's message for human beings.

"But the rest of you don't speak at all, do you?" Erna countered. "And meanwhile men and women insist the Indians are lazy by nature, because they only see the men adrift in the city, and look no further. Why should they? When what they've seen is convenient enough an excuse to withdraw all sympathy." She shook her head. "Things have gotten worse since the war. I'd never heard of anyone refusing to hire an Indian before 1918. Refusing to hire an Indian who is a Wobbly, sure, but that's something different. Now there's a cannery in Bellingham that only hires white workers. That's a hard thing for an Indian who served in the war to understand."

"It's amazing any of the men were willing to fight for this country in the first place," Will said.

"Some refused to go," Erna replied. "One of the Lummi men got in trouble with the draft registrars because he refused to abandon his parents' treaty promise not to engage in war. Yet they're the ones called savages. That was about the same time the state barred them from fishing outside the reservation. Despite the treaties." She shook her head. "Can you imagine what a law against fishing means to these men? To their families?"*

Trevor spoke up then. "It's a complex issue. I've seen the Indians be pretty wasteful when harvesting oysters. We're trying to base the management of the state's wildlife on a more scientific basis, and we can't have entire sections cordoned off from the law and science."*

"The oyster populations didn't collapse under their care, Trevor," Erna answered. "They collapsed under ours."

Will was clearly entranced by this new friend. The following week he took her up on the offer to visit the Tulalip Reservation. He returned home with a stack of paperwork on various legal cases he'd promised to look over on tribe members' behalf. He had clearly found work to do, and spent much of the spring up north or traveling between Seattle and Olympia trying to get a grip on relevant legal precedents.

Ordinarily Martin would have been relieved his dad found a cause worth fighting for in their new home, but one afternoon, after Ben dragged him away from his specimens for an hour at the Faculty Club, he listened to rumors that Leslie was very distracted lately by a pretty anthropology coed. Martin feared his dad's fascination with Erna's work might be just the kind of sympathy a neglected, young intellectual mother might find quite comforting. But though his dad could fall head over heels in love with a cause, Martin had not known him to be constant to a woman. Not since Mary.

So, on the last day of spring term, as he packed for the summer session at the university's biological station on San Juan Island, and Will packed for another trip to the reservation, he risked a reminder that not everyone, not even radical Boasian anthropologists, could adopt the theory that monogamy was an antiquated cultural practice and repudiate it in practice without suffering greatly.*

His dad replied with obvious irritation, "Good Lord, Marty. You were born in the wrong age."

"I just think you should be a bit cautious, Dad."

"And when did caution ever move the world?"

"I'm not talking about the world. I'm talking about you and the harm you might cause if you aren't careful."

"I'll take the risk if it saves me from a life of hesitation and uncertainty." He stuffed some shirts in a bag with a deliberate, angry movement.

"You live your life as you like, Marty. I don't interfere when you're going about things all wrong. I'm pretty sure you overintellectualized everything with Sarah, but I kept my mouth shut."

"Would you stop making this about me? And just stop and consider before you plunge?"

"God damn it, Martin. If I needed your advice, I'd ask for it."

He stalked off with his bag. Martin stared at his suitcase. The rare frustration in his dad's voice and the pointed reference to Sarah hurt. He had a vague sense that the last time he and his father had a fight it concerned Sarah. The argument must not have lasted long, for he had no memory of awkward, tension-filled silences between them. But they'd have one now, given he'd be miles away on an island for the next eight weeks.

Chapter 3

The first morning of the 1924 summer session at the Puget Sound Biological Station did not bode well. A steady June rain had fallen all night and not every canvas tent perched on the edge of the woods above Friday Harbor had held tight against the rain. Martin had fallen in love with the sounds and smells of these northwest rains, but the students came to breakfast with damp clothes and spirits. At sunrise Martin walked to the dining hall with Trevor, who had put together a full schedule and intended for work to proceed rain or shine. They both moved with the quick steps of naturalists used to the rain, but passed several groups of sullen, soaked students.

"I've got a plan to cheer them up," Trevor whispered with a wink.

A hearty breakfast was the first part of Trevor's plan. He enacted part 2 as the students finished their meal by rapping on his metal coffee cup and giving a short speech.

"It is exactly twenty years this month since we came to Friday Harbor to organize a marine station. Today you complain about a bit of rain. Try to put yourselves in our place in 1904 when our only laboratory was a table under a tree; our only shelter a dilapidated shack with no window and no door. We did our cooking over a beach fire. I assure you we didn't eat roast beef and apple pie prepared by university dietitians as you are doing now. But we had a dredge and the sea was full of strange animals and we knew the sun would shine tomorrow. We were happy as kings and rich as Croesus, not in money, but in hope and ambition. Today the sea is still full of animals and we can still go dredging. Now you go to your tents, put on warm clothes and heavy boots, bring blankets to wrap up in and we will go collecting. Biologists can't be softies and Washingtonians never worry about a little rain."*

That did the trick. Trevor's words cast the adventure in an entirely different light, turning physical discomfort into a virtue. Martin knew such sermons well: the great herpetologist Albert Gunther exclaiming what did he care for damp, dark quarters at the British Museum when thousands of unnamed snakes awaited him on the shelves. Or the entomologist Karl Jordan insisting he didn't mind working in a cramped, cold room as long as it had a large window so he had enough light for his microscope. Skillfully indoctrinated in this scientific asceticism, the students dispersed to prepare for the morning's field trips. Within an hour Trevor's morphology class had all cheerfully scrambled onto the station's research vessel, the *Medea*, to study tide pools on nearby islands. Hours later, disheveled from the wind and work, they marched back onto the station's dock, singing at the top of their lungs.*

> Oh we went to Friday Harbor
> Just to work a little harder
> Singing paddle, paddle, paddle all the way
> The natives shrug and call us Bugs
> Singing paddle, paddle all the way

Their boots were caked with mud and a few of the coeds still wore their trousers rolled up to the knees.

Having dumped their biological loot from the day's dredging on the tables in one of the laboratories, they cleaned up as much as possible and then descended on the station's dining hall, where large paned windows framed an impressive stone fireplace and provided a view over the harbor.

"Well done, ladies and gentlemen!" Trevor announced. He was still as disheveled as when he had disembarked from the boat. "At the moment the result of the day's dredging all seems a huge jigsaw puzzle you've dumped out on the tables, but by the end of the summer the basic pattern and plan of all life will be clear. You will see that the same principles operate in a starfish, a clam, or a man: feeding, eliminating waste, respiration, reproduction."

Martin, sitting near the fireplace and listening with the station's insect inventory in hand, was always amazed by how Trevor could make a statement that would have sounded sacrilegious to many Americans seem positively pious. Trevor got away with a lot, perhaps because he had such an absent-minded way of moving through the world. More than once he walked to campus wearing Louise's hat because he could not find his own. Although Louise always sent him out of the house on Monday with his pants neatly pressed and clean, by Thursday the cuffs were wrinkled and dirty and he was wearing one brown shoe and one black, all casualties of his constant search for interesting specimens on his way to and from class, deep mud puddles or no.

Given their experience with Trevor and Ben, the students on campus hadn't known quite what to make of Martin at first. They had no experience with a zoology professor whose desk was tightly organized, despite the chaos of specimen cabinets around him, and whose suit, though somewhat worn, was not disheveled by the end of the workday. But they were all a bit delighted by how he transformed into a "real naturalist" out here in the field. He still spent any spare moment trying to bring some order to the station's small collection room, but out in the woods and on the shore he rolled up his sleeves, didn't mind the dirt under his nails, and matched Trevor mile for mile on hikes to good collecting spots, whether it was pouring rain or sweltering. Though he never followed Trevor's example and abandoned his tie.

After that first field trip, rain or shine the students embarked on the *Medea* for a day's dredging, walked the shore for hours with Trevor, or trekked through the woods surrounding the station with Martin looking for insects. A few students who preferred lab work helped Ben with some experiments on the ability of various marine invertebrates to live in diluted seawater. When he wasn't in the field, Martin had a haphazard collection of insects to arrange in a room set aside in one of the station's two lab buildings. Old cabinets, a few chairs, and a desk near the window competed for space with a plant press, stacks of specimen trays, and piles of natural history journals.

Within the session's first week, a graduate student named Peter Harrison approached him after a day in the field and asked if he needed any help bringing the room to some order.

Martin laughed. "An army of taxonomists would be great, but I'm sure you'll do."

There really wasn't enough space for two human beings in the room, but within a few days a third joined them. After suffering with seasickness for three mornings on the *Medea*, a quiet botany student named Connor Huddleson asked permission to stay on dry land as well. So Martin and Pete moved some of the old cabinets into the hallway to make space for a table on which Connor could press plant specimens. Normally Martin would have hesitated at the thought of assistants, given the endless hours of training required to do good specimen work, but both young men had spent two years under Trevor's tutelage and had paid attention.

Pete tended to sit on the floor, his blond head bent over half a dozen specimen trays, as dark-haired Connor worked away at the plant press. Pete was tall and inclined to take breaks to chat now and then, while Connor was short and could work for hours in silence. Both seemed to love the work before them.

When Pete took breaks he had a habit of skimming a stack of old entomological journals, often pausing to read a particularly amusing gem aloud. One morning, as Martin sat at his microscope at the desk crammed between the cabinets by the window, and Connor bent over the plant press, Pete read them a poem from *The Entomologist*:

> When the Lord, the God of Nature,
> Had created every creature,
> These in Adam's view he placed,
> Who should name them to his taste;

And what he called both great and small,
Henceforth should be the names of all.

Adam soon his task began,
Named the beasts in wood that ran,
Birds that fly, and fish that swim;
Easy was the work to him;
And all he classed that met his view,
In genera and species too.

When at length his work was done,
He gave their names to every one,
And his system, as I hear,
Was simple, natural, and clear.
Thus without much toil, was he
Founder of Zoology.

But when he to the insects came,
His flagging energy grew lame;
So many species, and so small!
It was no joke to name them all;
And Adam said, "I'll do no more;
I think I'll leave the rest to Noah."

What creeps and flies was on the list
Of Noah, the Entomologist;
The Diptera, Orthoptera,
Hemiptera, Neuroptera,
The butterflies and beetles all,
And nothing was for Noah too small.

But the Ichneumons, as he found,
Were like a sea without a bound;

"This group is too confused," said he;
"I find it much too hard for me;
How to determine this mass well,
Not Cuvier could, nor Linné tell."*

Pete stopped reading, though the poem went on for several stanzas more.

"Don't the ichneumons lay their eggs in the bodies of caterpillars, and the larvae consume their hosts alive?" he said. "Not a very pleasant group to study. I'd say Noah had good reason to set them aside."

"Maybe," Martin replied without looking up from his microscope. "If he was paying any attention to the insect's behavior. But most taxonomy's done on dead specimens. He might have maintained a pretty blissful ignorance of what they do when alive."

"Still," Pete added. "Think a taxonomist's view of God depends on which organisms he studies? Butterflies give a pretty different portrait of the wonders of creation than these beasts."

"I suspect Noah and Linnaeus both knew how dreadful nature can be," Martin said. "Whether they knew about the ichneumon wasps or not."

"But Linnaeus attributed things like the existence of predators to the divine balance of nature. Goodness and wisdom of God and all that. But what possible good can larva slowly eating an animal alive serve?" Pete paused before adding, "Charles Darwin cited them as evidence against a benevolent God. Damning little facts, aren't they?"*

Martin finally looked up.

"Trevor found a species of ichneumon in Japan that helped slow the gypsy moth outbreak. I know an entomologist back east who sees those wasps as God's means of reestablishing the balance of nature when an insect oversteps its proper bounds."*

"Ah, but surely that's man's wisdom, power, and goodness," Pete countered.

"Or God's foresight in giving man the ability to fix things when they go wrong," Martin shrugged, turning back to his microscope. "Take your pick."

He did not catch Connor and Pete exchange a smile. Martin had become somewhat notorious for that shrug. Students witnessed it whenever they tried to make grand pronouncements about anything. Depending on their temperament and mood, they alternately reveled in or cursed its constant accompaniment: "Maybe you're right. Keep looking."

They had worked in silence for an hour when loud footsteps echoed on the boardwalk and Ben stalked into the room. He had set up two dozen aquariums in the other lab building. It was closer to the shore for easy access to the sea, with tanks filled with water of different salinity. His face was red, brightened by the fact he had walked into town to retrieve the mail and, it appeared, ran back. He threw himself into the one chair not piled high with specimen trays and let out a groan.

"Look at this." He waved a paper at them. "Some of those big-wig New York biologists convened a committee after I insisted they do something about Bryan. It's issued its first statement."

Martin had long since determined that the best way to get Ben out of the collection room so they could get back to work was to let him release some steam. So he took the paper in hand and read the names of biologists Edwin Conklin, Henry Fairfield Osborn, and Charles Davenport at the statement's head.*

"Bunch of phonus bolonus about the harmony between evolution and the Bible," Ben was saying. "I'm telling you. It's enough to make a man prefer *Hell and the High Schools*."

Ben had read excerpts from that book a few months earlier during an afternoon at the Faculty Club. It had made Bryan seem like scientists' ally. Bryan didn't call biologists sissies and soul-murderers. Bryan didn't insist the Germans who poisoned wells in Belgium and Northern France were angels compared to evolutionists who fed the poison of Darwinism into schools and led children straight to hell at taxpayers' expense.*

"At least the man sees evolution for what it is," Ben was saying.

"What's that?" Martin said as he handed the paper back, pretty sure he knew the answer already. He had heard Ben on the subject often enough.

"Completely incompatible with Christian doctrine, that's what," Ben answered. "Should be obvious to any man with guts enough to see the truth."

Martin shrugged and turned back to his microscope, which only made Ben more adamant.

"The cause of science is in deplorable straits when it must be defended by so-called scientists who attempt to reconcile it with primitive folklore. These are men with access to the *New York Times*, completely ignoring the fact science and religion are in direct and absolute conflict!"*

"So that's been decided at last?"

"It was decided at the beginning," Ben replied firmly. "But things are coming to a head, Sullivan. Politicians are soaking up Bryan's speeches and trying to put laws against teaching evolution on the books. We're going to lose an entire generation to superstition and ignorance if we don't do something! But I can't get Trevor to look up from his damned oyster experiments long enough to help keep this rubbish out of our state and defend human progress!"

Normally Martin did not reply to Ben's outbursts. But the one thing that could inspire him to risk the adrenaline and speak at length was when someone criticized a man who was not in the room to defend himself. He looked up from his microscope.

"That's not fair," he said firmly. "Trevor does his own part to help humanity. He burns plenty of midnight oil trying to save the oyster industry. That he refuses to become involved in debates that can't be decided by observation and experiment is his own business."*

"Ah yes," Ben countered. "But that stance might just be evidence he doesn't have the courage of his convictions."*

Martin crossed his arms and leaned back in his chair.

"Why don't you keep your crosshairs on me, since I'm actually sitting here? I wager I'm much worse than he is."

Though they had worked together for weeks now, Pete and Connor had never heard this tone of voice from Martin before and exchanged glances once more. But Ben didn't notice the change.

"And it's a damned shame. We need every man of science in this fight.

Fundamentalists are comparing biology education with Bolshevism, and accusing us of poisoning their kids."

"All the more reason to just do good science, and avoid getting dragged through the mud," Martin countered.

"Horsefeathers!" Ben scoffed. "Our entire field needs a good dose of Bryan's crusading instincts. We've been numbed into inaction by all this stay-above-the-fray nonsense." He leaned over Martin's desk to emphasize the point. "There's too much at stake for us to hide behind our lab desks and let the newspapermen do all the work. We've a responsibility to tell the layman what evolution means to humanity. And that it is accepted by all scientists."

"Assuming that fact carries much weight," Martin said. "I'm not sure it always does these days. Might just sound like propaganda from the papacy."

"Generally we're on the side of the Galileos, not the popes," Ben countered.

"As we see it, of course," Martin replied.

Ben shook his head. "Whose side are you on?"

Martin smiled. "I thought I was on the side of science. I'm not sure we know enough to tell anyone what evolution means."

Ben sniffed. "Christ, Sullivan. You're like some medieval monk. You think trying to do anything beyond fact gathering means you're giving in to the temptation of the world, the flesh, and the devil. But a bunch of scientists who just sit around gathering facts are of very little use to anyone."*

"I'd rather get a few facts right than risk being wrong about what evolution means to humanity."

"Those facts aren't going to matter a whit if the man on the street votes biology out of schools," Ben countered. "If we hide in our labs or museums out of fear we may misstep, then a lot of thoughtful men, thoughtful men with money, mind, are going to say good riddance and be done with us."*

"Careful, Ben." It was Trevor's voice. He had such a soft step that no one had noticed the sound of his boots on the boardwalk. He stood in the doorway, even more disheveled than usual. "You're going to convince Sullivan that he best leave off teaching and take refuge at one of Howard's experimental stations."

"He can do so if he likes but ultimately it'll be no refuge," Ben replied. "Not if Bryan succeeds in gutting biology education in this country. And he'll do it, too, if we all persist in just keeping our heads down. Sullivan can sit there with his big bag of depends, maybes, and who can say. But they'll get the university appropriations stopped if Bryan and his daft allies don't first."

Trevor laughed, which was how he usually dealt with Ben.

"You need to move to Oklahoma, Cardiff. They passed a law against teaching evolution last year. You'd have a good fight there."

"It'll arrive here soon enough," Ben muttered.

"I doubt it," Trevor said, placing some newly filled specimen vials on Martin's desk. "We haven't had trouble out here in decades. An old minister complained about us in the 1890s. City's bootleggers and rising hemlines have given 'em bigger fish to fry these days."

Martin used the excuse of Trevor's arrival to look through the lens of his microscope again. He could see very little, as he tried to calm the adrenaline it had cost him to defend Trevor. Pete had set aside his work and looked at Trevor.

"You ever have a student who didn't believe in evolution?"

"Oh sure," Trevor answered. "Years ago. A serious young man warned me at the beginning of my evolution course that he didn't believe in Darwinism. Said he was there to see what arguments I could concoct in its favor. Attended class every day and took careful notes. Passed the final exam just fine. And added a postscript that I put up a good bluff, but I'd have to find better arguments to convince him that a man and a monkey are brothers. Still. He came to a considered decision on the matter."*

"How the hell could that have been a considered decision?" Ben said. "If he graduated from this university believing in a damned fable? No man should be allowed to pass a science course if he denies the foundation of all biology. When Darwin destroyed the boundary between man and animal, biology became the only rational means of understanding ourselves. Of all progressive action. Of making better citizens. If this generation doesn't understand that they must be guided by biological facts rather than fairy tales, then they'll continue to be led by emotional ignorance. They'll sanction acts that in the end lead to social disaster."*

"Just not sure how well we've mapped it all, yet, Ben," Trevor replied with a slight smile.

"The churches have claimed to have it figured out based on a bunch of fables. Anything we've got is better than that. The sooner bishops and ministers join skirts that reach to the ground and go the way of the Victorian age, the better."

"Hold on now, Cardiff," Trevor said with a more serious tone. He tended to use last names when he was either teasing or upset. Now, he was upset. "Churches perform very valuable functions. They are responsible for much of the progress of civilization. As a means of organizing and unifying society, they have few parallels."

Ben scoffed, but before he could reply, they were all a bit surprised by Connor's quiet voice.

"But that's not admitting the validity of its content. Only that the church is useful, true or not, and might as well go away now that its work is done."

Connor was usually so silent that Trevor, who could be as evasive about such things as Martin, replied at greater length than usual. Even though Ben was in the room.

"Well, yes, I'm not sure we need the buildings and paid intermediaries between man and God anymore. But giving those things up doesn't mean we must give up what made religion a useful adaptation in the first place. My own view is that the beautiful, natural world unspoiled by man is the best cathedral. Surely the daily rising and setting of the sun, the regular passage of the seasons, the eternal rise and fall of the tides, and the endless progression of the stars in their paths, prove that such perfect coordination of nature's laws could not have been achieved without a master plan behind it. Better than any sermon in a church."

Even Martin paused his work at Trevor's little speech. He'd never heard his boss talk like this.

Connor was watching Trevor closely. "But that isn't Christianity."

"No. It isn't. Not according to most definitions. The Christian miracles, the promises of Revelation, the idea of a life after death, an immortal soul. They're all contrary to facts. We know that man was derived through a complex evolutionary process from an anthropoid over millions of years. Well then. If a man has a soul, how did it originate? Did that first ape-man have a soul when it emerged? Did it exist in the sperm, in the ovum, in the fertilized egg, in the embryo, in the gastrula, or in the squalling infant when born? But I'll tell you what I do know for certain. I know that the entire cosmos from the giant sun to a single atom of hydrogen is ruled by universal chemico-physical laws. That is the God of the scientist. Awe-inspiring. Inscrutable."*

"Good grief. You sound like a minister," Ben grumbled at last.

"Have you noticed that when we're working in the lab and the room is silent, we might as well be Trappist monks?" Trevor said. "Some laboratories even have signs that say no talking all over their walls. Now where do you suppose we got that idea? The first scientists were monks, Ben. I wouldn't be too hard on the vestiges." He looked at Connor, and the mischievousness gave way to sympathy. "Darwin felt that the subject of God is too profound for the human intellect. He compared trying to understand such matters to a dog trying to imagine what a man was thinking. I don't remember the words exactly."*

"A dog might as well speculate on the mind of Newton. Let each man hope and believe what he can," Martin recited without looking up from his microscope.

Ben threw up his hands in mock disgust.

"Well we've obviously got enough sermonizing going on around here."

Trevor laughed. "I think that's actually a refusal to give a sermon. Sermons can lead to crusades, you know. And I don't think Martin's got a crusading bone in his body."

Ben smiled. "Indeed. I suspect that may be a source of great disappointment to certain young ladies on campus."

Martin was used to Ben's ribbing, but the sudden change in topics caught him off guard, and he looked up in confusion.

"You're a fine taxonomist, Sullivan, I'll give you that," Ben explained. "But you're damned oblivious when it comes to certain organisms. Rumor has it you've already let some pretty specimens escape your net."

Trevor joined the interrogation with a laugh.

"Well, Dr. Sullivan. Defend yourself."

"I'm afraid my case has already been decided," Martin replied. "I'm obviously a bad collector of something. But I never observed any indication I'd be warranted in trying to set a trap."

"Were the taxonomist to focus only on the most conspicuous characters, where would he be?" Trevor winked. "Someone complained to Darwin once that it was difficult to get a definite answer to a simple question from Nature. He replied, 'She will tell you a direct lie if she can.' You've got to have your wits about you, Sullivan, if you're going to get the truth out of Nature or a woman."*

Martin frowned. "And yet I am supposed to have diagnosed the willingness of one to be captured?"

"Ah, but it isn't so difficult," Trevor replied, "if it tends toward our own desires."

Martin's frown turned into a slight smile. "Isn't that the kind of diagnosis we're supposed to avoid?"

"Good lord, you overthink things." Ben laughed. "Jack London insists love is simply a mad, happy delusion. Just sexual desire dressed up in a veil of civilized fairy tales. No use thinking about it too much."*

Trevor paused and his brow furrowed.

"But Sullivan's got a point! If selfishness, desires, and emotion are the remnants of our animal ancestry, then aren't we supposed to, as human beings, do everything in our power to vanquish them? Isn't that what science is all about? It's a dilemma! Our new brain that confers reflection and restraint, the imperious command to stop and consider, is in conflict with millions of years of evolution, yet the source of all civilization. But how do we know when our old animal brain is driving us? And surely there are times when impulses, our old brain, must be trusted in the interest of survival. Life-or-death crises. Love. Sooner or later, all doubt must be at an end, and a decision made."*

Martin smiled at his boss's enthusiastic speech. "And if we give in too soon?" he said.

Trevor smiled too. "Ah. Well. I suspect a smart young lady will point it out to us. That tends to be my own experience."

Chapter 4

The Puget Sound Biological Station's summer session was in its third week when some of the students asked if they could walk to a rocky outcrop christened Heaven to watch the sunset.

"If you can find someone to go with you," Trevor replied. "I think it's Professor Sullivan's turn."

Trevor was no prude. He let the students play jazz on the old upright piano in the dining hall, though he had colleagues who argued jazz was leading to an epidemic of physical and moral degeneration. "Just as likely it's modern living and machine civilization that's perverting human beings," he would say. "Who's to say the music isn't restoring a bit of rhythm and emotion to life?" He turned a blind eye to occasional cheek-to-cheek dancing, though it was a style banned on campus for fear it would spoil the university's good name among those in charge of appropriations in Olympia. But he insisted that an elder—Fire Extinguishers the students called them—accompany these sunset excursions. Martin had at first balked at the idea he serve as chaperone. Ben might call him a monk for his strict rules regarding what a scientist should do with his time, but he hated the idea that he must police any other behavior of the young people in his charge.*

"Is this part of my job description?" he asked Trevor.

"It certainly is," Trevor replied with a smile. "Comes with the territory out here. Don't worry, Sullivan. I doubt you'll have to say a word. That's really why the ministers are nervous, I wager. We're competition."

"The rules don't always change much, though, do they?" Martin replied.

"Doubt Bryan sees it that way." Trevor laughed.

It was a treacherous walk at points. The path led along a steep, eroding bluff above the rocky shore. Martin asked two of the coeds to change out of the small-heeled shoes they'd worn to dinner and into sturdier boots before embarking, and then led the group to the trail. When they came to the clearing high above the shore, he sat apart from the students as they waited for the sky and landscape before them to change. Trevor was right. He did not have to speak a word of reproach to the students sitting about him as the sun set.

A full moon lit up the water once the sun dropped behind the mountains and Martin was just about to signal it was time to return to the station when one of the students moved to sit near him. Her name was Phoebe Bartlett, and she had been sitting with Pete and Connor a short distance away. Martin had never had Phoebe in class but she sometimes stopped by the collection room. He had noticed that although she was better in the field than any of them, some of the male students tended to avoid her. She had an uncanny ability to be completely silent and still when observing nonhuman creatures, but extended neither skill to witnessing certain actions of her fellow human beings. He had heard she sent a letter to the student newspaper protesting the dean of women's list of potential careers for coeds because it was limited to home economics, teaching, and secretarial work. Martin suspected she made the young men uncomfortable by applying their freethinking ways to how they should view and treat coeds. She was friends with Pete and Connor, though, and Martin thought that was why she often visited the collection room. He was glad it was too dark for her to see the blood rush to his face as he remembered Ben's teasing. He wondered if Ben had noticed her visits too, and misdiagnosed them. One of them must have.

"Pete told me you did some collection work at the museum in New York," she said now.

Martin nodded. "Off and on, in the summers. They have a good collection of Anthribidae from South America."

"Ever meet some of the men at Columbia? The geneticists, I mean?"

"They didn't really need us much," Martin replied. "But I went to some of their seminars."

"What are they like?"

"Lots of graphs."

"To work with, I mean."

Martin's brow furrowed. "They work hard. Some call them single-minded, but the tactic gets results. It's a good sight cheaper than collecting beetles and butterflies. Why?"

"I've been offered a position as a lab assistant for a year in Thomas Hunt Morgan's lab. Professor Kincaid said I best put my hat in the ring for as many posts as I can. That's code, perhaps, for the fact most aren't likely to take a woman. I'm not sure he had Morgan's fly lab in mind, but I hear some of the labs are more likely to hire women. Of course, the old naturalists will think I'm selling my soul if I take the post."*

"Don't listen," Martin said firmly. "There's nothing wrong with lab work, if it's done well. Morgan can breed thousands of flies in a week and track inheritance in real time. He's carved out a corner where he can just focus on getting good science done, no matter what's going on in the rest of the world."

"Gracious, you sound jealous!"

"Of a man tied to a lab bench? No thank you," he replied. "But there's no harm in learning both kinds of work. There are certain things we can't do with dead specimens, especially now. If we ever figure out how evolution works, it's going to be through combining the best of both methods. They can control their little organisms, but they sometimes forget the complexity. We can't control our little subjects, unless killing them counts, but we tend to remember the complexity." He paused for a moment, and then added, "But I thought you preferred fieldwork."*

"Preference isn't enough to make a career out of something, is it?" she replied. "Imagine I preferred taxonomy. You worked at Harvard. Exactly how many women are on that museum's staff?"

Her question sounded rhetorical, but he answered anyway. "A few. I don't know how many exactly."

She laughed. "I hope you aren't as poor an observer of your insects as you are of the humanity about you."

There was a rather abrupt candidness in the words that made Martin smile.

"I suspect there's a direct inverse correlation." He paused, mining his memory. "There were a few artists. And the librarian."

"And how many entomologists?" Phoebe pressed. "Curators. Paid experts. That kind of thing."

Martin hesitated. Elizabeth Bryant had worked on spiders at the museum for decades, but though Sam had supported her in a liberal manner, compared to those who viewed a woman in the museum as a distraction, he had never extended that liberality to actually paying Miss Bryant for her work.*

"Not on the payroll," he said finally.

"Exactly. Women can write specimen labels, tidy the library, and write checks if they're wealthy. But they aren't usually hired for scientific positions. No matter how good they are. Most men will put my application in the trash bin on the grounds I'll most likely marry and it'll all come to naught."

Martin felt any reply to that remark would sound trite or defensive, so he said nothing. He knew very well she was right.

"Hard to argue against them when physiology implies that interfering with the way of things is immoral. Do you know what Professor Cardiff teaches in his classes?"

Martin frowned. "Experimental physiology, I assume."

"I suppose one could call it that," she said. "He's been assigning a bunch of stuff from Geddes and Thomson's *The Evolution of Sex.* They start with the biology of protozoa and end with man. And what's the grand conclusion of it all? The duty of homemaking must fall on the mother because women are designed to provide good environments for the strongest and wisest men. I thought no thinking biologist could claim such things in 1924. It's tempting to just abandon science altogether. When it's just being turned into a new church through which half the human race is told to keep their proper place and be quiet."*

Martin did not like speaking about Ben when he wasn't present, as much as he despised the confident generalizations Phoebe described.

"Have you spoke to Professor Cardiff about this?" he asked after a long pause. "Sometimes he listens well enough."

"Oh, that's fine, coming from you, professor," she answered. "You who sit in silence when he starts talking about the survival of the fittest."

Martin smiled. "Fair enough. I just have a hard time joining in when zoology professors argue about things we know very little about."

"But there's a great deal at stake. For some of us," came the frank reply. "You have a legendary sight about you, but you are blind about certain things."

He laughed outright at that. "I don't doubt it."

"Hey professor!" It was Pete's voice, a few meters away. "Think the Nereis worms will be convening for a dance tonight?"

Trying to pay attention to Phoebe's concerns, Martin had forgotten all about the moonlight and what it might do to a certain species of worm that lived beneath the sea. He hollered for everyone to head back. When they arrived at the station buildings, Trevor and the students were running about gathering dip nets. For weeks Trevor had been anticipating the first full moon of the summer, when two-inch, segmented worms that lived out their days in the sand might wriggle free of their safe enclosures to reproduce. Soon twenty students and their professors were kneeling, dip nets and battery jars in hand, all along the station dock's edge in silence, watching and waiting.

"The pale green ones are the females," Trevor explained. "There's one. Just there."

They watched the lone female for a moment. Then Phoebe saw more and wondered aloud, "Aren't the males coming to the party?"

She knelt a few students away from Martin, leaning over the water, her eyes fixed on the drama below. Like the other coeds, Phoebe's light brown hair was bobbed, but so thick that it kept falling over her ears so that every few moments, very slowly, she would press it back behind her ear with her fingers. Pete was kneeling beside her.

Everyone waited to see if the males would, indeed, "come to the party." And within a few moments they appeared, swimming slowly around. The water gradually changed to a white blur.

"The males are releasing their sperm," Trevor explained.

As the worms moved about, their eggs and sperm mixed outside of themselves in a dance of polygamous and polyandrous promiscuity.

"The females are dropping their eggs," added Louise, kneeling beside Trevor.

Soon both sexes dashed through the water in a frenzy. The students were silent as they watched the worms. There was not even one off-color joke, which tended to be a specialty of some of the male students and one or two of the coeds. After half an hour the worms' behavior changed. Both sexes became listless, and one by one the individuals disappeared into the depths. After a few moments, the sea was once more still and blackish green beneath the lights. The frenzy of reproduction ended with every individual's death.

"Too bad they had to die," Phoebe said quietly, and then added with a hint of sarcasm, "But their mission in life has been fulfilled, I suppose."

The students collected some of the eggs and sperm in the battery jars and carried them back to the lab. They examined their treasures under the microscopes in the hopes of witnessing fertilization and the first divisions of the egg cells, diligently sketching each change in their notebooks. Martin, Trevor, and Ben wandered from desk to desk to make sure they were watching eggs and not specks of dirt on their lenses. More than once Martin overhead Trevor say quietly after a student gasped in delight, "It is a moving experience to observe."

When the students all seemed comfortable at their microscopes, Martin found an instrument that was not being used and for hours watched his little stash of eggs. Each one was slowly bumped by a sperm into the state of a zygote, which then divided into two, four, sixteen, and so on, until, eventually, something that passed for an organism appeared. He thought of the millions of eggs out in the water, some ever so slightly varying as they divided, lined up from weak to strong (or was it the bad luck of a wayward current?).

Phoebe's interrogation at the cliffs reminded him of Trevor's words, a few weeks earlier, urging students to see that feeding, eliminating waste, respiration, and reproduction were the basic principles uniting all organisms, from starfish to human beings. As a naturalist, or biologist, as the younger set called him, he was expected to lecture about reproduction. He had once read in a biology textbook that "the vague sexual attraction of the lowest organisms has been evolved into a definite reproductive impulse, into a desire often predominating over even that of self-preservation; that this passes by a gentle gradient into the love of the highest animals, and of the average human individual." In abandoning the barrier between animal and man, biologists had set their sights on biological explanations of human affection, collapsed romance into animal courtship, turned human love into the sex instinct, and made science an authority on both. But where the biologist must pass the topic to the philosopher, poet, or minister never seemed clear.*

"Like worshippers at prayer," Trevor whispered, leaning over him. Martin looked up at the room and smiled. Trevor was right. The room had fallen absolutely silent.*

Until Ben became bored of just watching development and began brainstorming with his graduate student Eugene Sanders about how one might intervene in the process to actually figure out what was going on. As often happened, Ben's conversation eventually turned into a discussion of how biology should be reformed. Martin happened to be at the microscope nearest Ben and Eugene, and might have escaped the ensuing debate by helping students on the opposite side of the room. But that would have meant abandoning his own drawings of the embryos. And he had never witnessed Nereis worm development before. So he stayed put and tried to ignore Ben's animated manifestos.

"Too many biologists insist their work has nothing to do with human struggles," he was saying. "But it's absurd to take endless pains to understand the behavior of sea anemones and earthworms and leave man alone. Not enough Americans understand that creatures like this provide clues to understanding themselves."*

Trevor heard Ben's words as he passed, and laughed. "Are you talking about war or mate choice?"

"Both," Ben replied without hesitating.

"Well then," Trevor said. "All sounds fine if we've figured out human nature."

"Oh yes. And that sounds quite cautious," Ben said. His tone clearly indicated he did not consider caution a virtue. "And damned immobilizing. It leaves explanations of human behavior to utopians and clergymen who imagine man can be a different being than he is. If we aren't willing to tell the man on the street what we can learn about his drives based on his ancestors, then we leave the organization of society and control of men to those who believe man is made in the image of God. Did you know there's a judge in Chicago who's got down off his bench and is studying biology and psychology in order to understand the criminals he was hanging? That's courage. He's discovered that biological psychology acknowledges that men are motivated, not by logic, but emotions, wills to power, and animal drives and desires, with an extraordinary capacity to explain all his irrational actions with what he thinks are rational arguments. No matter what we think man ought to be."*

Trevor nodded. "Yes. But sometimes I wonder whether a primary means of improving mankind might be the simple act of imagining he's better than he actually is. You know. Take the imaginative leap, that maybe our animal instincts can withstand the demands of the League of Nations after all. Or pacifists."

"Come on, Trevor. How's the pacifist going to help us? He asks us to believe that men hate war. Talk about a bunch of biological buncombe. We've got to legislate for a human being that exists, not for one who does not. Men love war. They always have, always will. It's a matter of basic physiology. Hormones. Grand ideals about peace and cooperation are no match for the endocrine system."

"Well, you may be right," Trevor said. "But I hope not. Besides, the ability to imagine peace and cooperation must be just as biological as the desire for war."

"Ah, but our biology might not be able to deliver on the hope, that's my point."

Martin had kept his eyes on the embryo as Ben and Trevor talked, but suddenly he felt Ben leaning over his shoulder.

"And what does Professor Sullivan think?"

Martin did not look up from the lens of his microscope as he replied, "At the fourth cleavage the blastomeres are distributed in four micromeres divided unequally in a right-handed spiral, and the four macromeres divide unequally in a left handed spiral."

Someone laughed, and Martin looked up and saw Phoebe sitting at the microscope across from him. She was resting her head on her hand and smiling. Then she spoke.

"Can I ask you something, Professor Cardiff?"

"Shoot," Ben said.

As she spoke she glanced toward Martin with a challenge in her eyes, as though to say, "This is how it is done."

"Our textbook says that throughout the animal kingdom males are active and females passive. That this division of labor and physiology began with the protozoa, and must thus be directive throughout the animal kingdom, including the higher animals. Geddes and Thompson insist that what was decided among the prehistoric protozoa cannot be annulled by Act of Parliament. Do you believe that's what the study of protozoa tells us? Is that a proper extension of what we can learn from these little creatures to the troubles of humanity?"*

Having encouraged her to tell Ben of her concerns regarding the claims physiologists were making about sex differences, Martin was relieved when Ben took her question quite seriously.

"I have no doubt that a whole lot of nonsense about such things has been written in the name of biology," Ben replied carefully. "Some of my colleagues claim women are a degradation of the male type. I don't happen to think that's what biology tells us at all. But I do hold that there are real differences between the sexes that must be acknowledged if we are to truly know ourselves. After all, it's the job of biologists to classify and explain such differences. To do anything else is to take comfort in a delusion because we desire it."*

"But you do a great deal more than classify those differences," Phoebe countered. "You insist those differences tell us what a woman should be and do."

"When one considers the race as a whole, rather than the individual, yes, I insist that biology tells us that the work of women lies primarily with the young. If women understood that the home and schoolroom is as large a factor in human progress as the railway and the telegraph, maybe they'd understand that their importance is increased, rather than diminished, by taking the biological point of view."

"And what if a woman doesn't want to be a housewife or a school teacher? What then?" Phoebe pressed.

"I agree that not every woman, as matters are, can find occupation in household cares and in the training of children," Ben replied. "And to the extent that women aren't so occupied their need for thought and action is not essentially different from that of men. A woman, like a man, must find something to do if she is to avoid misery and decay."

"But surely finding what a woman should do in nature makes anyone who chooses differently a second-class citizen, or a pathological exception," Phoebe protested.

"No, no," Ben countered, shaking his head. "I don't think it does."

"But do you honestly think that these classifications are so rational, when there are profound differences among individuals within those groups? Differences that belie your generalizations?"

"I'd say you are giving up one danger, that of overgeneralizing, for another, the belief that our categories mean nothing outside the space of our own minds. In which case we might as well just sit and imagine what things are like, rather than do science. Right, Trevor?"

The appeal for aid made Trevor, sitting in the corner, release a short laugh.

"Well now, I'd say you need a good taxonomist to answer that," he said, gesturing to Martin. "They're the ones obsessed with whether or not our little categories are real or no."

Martin, who had given up trying to focus on the embryonic development taking place on the slide, only smiled slightly. He did not want to give the impression Phoebe needed a man's aid in debating anyone, and had long since resolved not to argue with Ben in front of the students. So in the face of his silence Trevor replied.

"I think Thomson and Geddes are, in theory, justified in their basic thesis that biology tells us men should fight if there's any fighting to be done, and that women should nurse if there's nursing to be done." He paused. "I'm not so sure about the tendency to extend the thesis to all other professions. Take medicine. I know some women who are doctors, not nurses. Darn good ones. Take that single observation, and one's conclusion must be qualified accordingly. Caution and more observation seem to be the main requirements of the case."

"But that's just it," Phoebe replied. She looked at Ben. "You don't have time for caution and more observation. Not when biologists are promising to save the world now by telling everyone what human nature is like and what we should all be doing."

A few of the undergraduates shifted uncomfortably. They were not used to a student, much less a coed, challenging a professor like this. Martin caught one of the male graduate students rolling his eyes. But Ben looked at Phoebe with a furrow to his brow that showed he was thinking over some puzzle she had presented, rather than annoyed. Before he could speak again, Trevor ordered the students to set their microscopes aside.

"Let's all call it a day," he said. "It's almost midnight. From this point the development slows down."

Martin glanced at Phoebe as she stood up and started putting equipment away. She smiled and gave an exaggerated shrug. Then Martin turned to see Ben watching them, smiling.

"You two in cahoots, are you?"

Phoebe laughed dismissively. "And when did Professor Sullivan ever say anything more dramatic than the cleavage of blastomeres?"

The next morning, after a large breakfast, the students put on their boots and hiked down to the tide pools. Normally Martin would have

worked in the collection room. He had dozens of newly collected specimens that needed to be pinned and identified. But Trevor had asked him
to be on hand to help students with their identifications. Since his first
visit to the island, he had spent as many hours learning the tide pool invertebrates as walking the woods to study insects, and Trevor had found
him out when he was the only one who could identify a rare echinoderm.

So as the warm sun washed over the black rocks and the students had
dispersed across the shore in little groups, he bent over a tide pool trying to focus on the little nonhuman animals in his shadow. Ben's student
Eugene walked up and asked for help identifying sea anemones. When
Martin found a few distinct specimens for him and described their key
differences, Eugene seemed quite uninterested in the challenge of learning such detailed distinctions. He had clearly paid little attention to the
character and temperament of some of his professors as well.

"Fine row last night, wasn't it," he said. "I wager Miss Bartlett's just
sore because she didn't get a fellowship at the Smithsonian's biological
station in Panama. They only go to men, you know. Doesn't make sense to
give it to a woman when the station doesn't have separate water closets."

Martin knew better than Eugene why that was. He'd worked alongside entomologists who had been to the station. The station in Panama
didn't have separate toilets for women because its main patron, David
Fairchild, had balked at the proposal that a women's dormitory be built.
The "sex element" would take over, he argued, and destroy the pure, intellectual sparring he loved. No, Fairchild insisted, let there be a place where
real research men can find freedom from such distractions.*

"She should take that position in Morgan's lab," Eugene was saying.
"He'll turn her into a good old-fashioned lab assistant. And if she doesn't
like it she should just get married. Or become a fieldworker for the Eugenics Record Office. 'Course they'll only hire her for three years since it's
best that the women smart enough for that kind of work get married."**

"And if a woman doesn't want to be a wife?" Martin asked, turning
over a rock and watching the crabs beneath scatter without actually seeing them.

"Don't you think that's a bit unnatural?" Eugene replied.

Martin laughed. "Spending one's days examining crabs under a rock isn't very natural either."

"But science is surely a cultural extension of man's natural need to understand his environment so that he may survive in it."

"Aren't we a bit degenerate, then, if we aren't all doing applied biology?"

Eugene was not amused. He took up Ben's cudgels in his absence.

"Professor Loeb says the instinct of caring for children is just as inherited as the morphological characters of the female body. What's a biologist supposed to think of a woman who shuns the duty of marriage and childbearing and pursues a life of science?"

Eugene was parroting claims Martin had heard often enough. Ben took great pride in having worked with the famous physiologist Jacques Loeb before the war, and often spoke, as Loeb did, of the instinct for food, sex, and care of one's young as resting on a chemical basis that would be subject to mechanistic analysis in future. But in marked contrast to Ben, Loeb argued that men could develop a universal ethics based on the knowledge of hereditary instincts like workmanship, maternal love, and a compulsion to see one's fellow beings happy. Martin didn't have the heart to tell Eugene that he'd seen Loeb after the war, after his once-confident claims seemed naïve or worse. Or that colleagues attributed his physical breakdown to the war's terrible evidence that describing humans as reflex machines made just as much sense of the ease with which men, including Loeb's fellow socialists, had been driven willingly onto the battlefields of Europe.*

"I don't know," Martin said instead. "You can cite Professor Loeb and a few others. Most of the experimental men keep silent when they move away from their lab desks, lest they spout nonsense."

It was a harsh thing to say and Martin regretted his word choice at once. But Eugene stalked off before he could come up with a better answer.

One of the few things Martin disliked about teaching was how, in moments like this, when he felt he had failed at whatever guidance he was supposed to be providing to the students in his charge, he could temporarily lose his ability to concentrate on his scientific work. It didn't happen very often, thanks to his strict rules about avoiding debates if they weren't resolvable with more specimens. But that evening, quite certain he wouldn't be able to concentrate after the exchange with Eugene, he didn't go to the collection room after dinner. Instead he sat by the fireside as Louise entertained little Marjorie nearby with her favorite wooden train set, and the students danced while someone played "What'll I Do" on the piano. Trevor sat next to Louise, immersed in the *Proceedings of the Biological Society of Washington*.

"You don't dance, Martin?" Louise asked as they watched the dancers. Ben was dancing with one of the prettiest coeds.

"Not much these days," Martin replied with a slight smile.

Louise smiled too. "Well. Perhaps it's just as well. Theresa McMahon says we should make a rule that our professors not, in fact, be allowed to go to dances until the coeds have a more realistic idea of their salaries. It can be quite a shock for a young lady with visions of a home with all the new cleaning gadgets and the latest fashions. A husband in love generally has to go into some applied line to keep the house up to standard. There wouldn't be much time left for your beloved weevils."*

Marjorie tugged on her skirt with a "Mama!" and Martin got a two-minute respite from the topic as Louise found various toys squirreled away in her handbag. When Marjorie was sufficiently entertained, and in the face of clear evidence via Ben's teasing that he was the subject of campus gossip, Martin tried to halt conjecture by at least one respected friend.

"I was almost engaged once," he said. "She was a chemistry student at Radcliffe. Married a minister instead."

"Gracious," Louise replied. "How did you lose her to a minister?"

"I don't really know. The memories are all very muddled. My mother was very sick at the time." He paused, but Louise waited in silence. "I met my rival once. He was a Unitarian, which is I guess pretty liberal, as far as ministers go. He insisted science can teach men and women how to keep our preferences and prejudices from coloring how we see the world. That the ministers needed us on their side, if Christianity is ever to be purged of intolerance and superstition."

Louise smiled again. "He lectured you."

Martin nodded. "Said we aren't that different from ministers. They try to escape the confines of their own minds in order to discover and obey God's purposes. We do so to try and understand nature rightly. He believes we're allies in the effort to convince men and women to look outside themselves in order to determine our duties and destiny."*

"And what did you say?"

"Probably something about how I'm an insect taxonomist and don't deal with man's duties or destiny."

"And your almost fiancée? Was she there?"

Martin nodded. "They got into an argument about whether science is the new church. And whether that's a good thing."

He did not mention the only other things he remembered: that he had never seen Sarah so animated as when she debated that earnest minister, and that he had been bewildered by the fact he felt no jealousy, an emotion that he was apparently supposed to feel if he was in love. Instead he said, "It's strange. I can remember that conversation well enough. But I don't remember how things ended. With Sarah." He looked up and smiled, as though in apology that he could not tell her more. "Very muddled, you see."

Neither had been speaking as though Trevor was part of the conversation. He had a talent for remaining unaware of his surroundings when reading. He surprised them both by weighing in without looking up from his journal.

"No doubt Freudians would say you've repressed a painful memory of a moment when your natural instincts struggled with conventional morality. Civilization won, and jealousy and conflict were repressed. Of course, they also insist you best get the memory back by paying a hefty fee to a psychoanalyst else you'll develop a stutter, tic, or something worse. Personally, I prefer Bergson's theory that the human brain is designed to restore individuals after great suffering. We think it is the organ of thinking, but it is the organ of forgetfulness. And that's the only reason we get through life without going mad."*

Louise shook her head.

"Nonsense. Perhaps their theories work fine for intense suffering. Shell shock, maybe. I hope they do. But surely in ordinary life we don't suppress emotions with such complex mechanisms." She turned to Martin. "You probably went home and got drunk, and that is why you don't remember more."

Martin stared at the fire. Louise was right. It all must have happened on the same night. The break up with Sarah. The fight with his dad. The bottle of whiskey.

She always came up to the apartment for tea after they attended some social event, like the dinner at which he met his rival. When his mother lay ill in another room, Sarah was the one visitor with whom Will would sit for longer than five minutes. Though she never imposed the sentiment on Martin, Sarah loved a good argument almost as much as Will did, and he always had a stack of clippings about current events or recent articles in the *Truth Seeker*, his favorite freethinker magazine, on hand to place before her. They argued about whether her stylish, bobbed hair was putting hairpin makers out of business, or whether the suffragettes' hunger strikes were good strategy for winning over congressmen to their cause, or whether women wore high heels to please men or because they liked the style. Will was the one who had started calling her Martin's fiancée, which was why the "almost" had slipped into his conversation with Louise.

Martin remembered now that it was on the first evening he returned from some dinner alone that he had yelled at his dad for his barrage of "Where's Sarah? You didn't let her get away, did you?" He remembered the look on his dad's face. He remembered his mother was gone. And he remembered that in the ensuing silence they had voicelessly agreed to split the single bottle of whiskey that had stood for years in the back of a cabinet for medicinal purposes. But that was all that came back. He realized that bottle of whiskey meant he might never remember his and Sarah's final words, or why they had parted. Biologists who supported prohibition would no doubt insist that he had impaired the fragile nerve cells on which memory depended by turning to alcohol. Some would even say, in alliance with teetotaling ministers, that he deserved the loss to boot.

He had once heard a famous biologist compare love to intoxication. Biology, the biologist announced, demonstrated there was more science in the phrase "love drunk" than most wished to acknowledge. Love was nothing more than a chemical trick, designed by natural selection to ensure the persistence of the species, and it captured the will just as much as alcohol did—perhaps their chemical pathways were even similar!—and inspired the most rational of men to foolishness.*

This equation of love with the will-altering incapacitation of drunkenness had never made sense to Martin. He had never felt the complete loss of self-control described in such biological accounts of love. As little Marjorie distracted Louise once more and he was left alone with his thoughts, he wondered whether drinking that bottle of whiskey with his dad had been a vain effort to test whether he could feel intoxicated, by love or alcohol, at all.

Part II

The Ministers

Chapter 5

The biological station's summer session was in its final days when some of the students asked Ben and Martin to join them for a pre-dawn walk to the cliffs to watch the play of the full moon on the water. Martin was bringing up the rear and just reaching the more treacherous parts of the trail, high above the shore, when shouts arose in front of him. Urgent cries of "Connor! Connor! Are you all right?!" were followed only by the monotonous sound of the waves gently lapping the rocks below.*

A few students ran back to the buildings for better lanterns and rope. When they returned the dawn light had just begun to illuminate the body lying on the rocks below. By that time Martin and Ben had risked life and limb to make their way down the steep cliff and reach the tide pools before the ropes arrived. Martin was at Connor's side first, for he could use skills reserved for retrieving rare specimens, while Ben, who spent his days at a lab bench, had to move more slowly.

Focused on where to put his feet and hands, Martin hadn't heard the sound of a motorboat passing close to shore as he descended. As he reached Connor he realized that the waves were much louder than normal, and the violent crest of a boat's wake would soon pull them both into the harbor. His hands became covered in blood as he tried to pull the body away from the water's edge. Ben made it down just before the largest wave hit and helped pull Connor completely out of the water.

"Damn rumrunners," Ben muttered.*

For a moment they both sat very still in silence. Then Ben pressed a hand on Martin's shoulder, for when there was no cliff to navigate and nothing to be done, the adrenaline had nowhere to go and Martin began shaking. Ben commanded him to sit back against a boulder, and, cupping the cold seawater in his hands, cleaned Martin's fingers.

The next twenty-four hours were a blur. Someone must have corralled the students and led them back to the lab, for eventually all Martin and Ben could hear were the waves coming in and out of the tide pools around them. They did not speak. For a long time, Ben sat with his head resting in his hands and his eyes closed. Martin stared at the waves as the scent of decaying organic matter that he had grown to love along these shores turned his stomach for an hour. He had stopped shaking but it took all his concentration to prevent himself from vomiting as the early morning light transformed Connor's shadowy form into a corpse.

When Trevor appeared, coming toward them in a small dingy manned by Friday Harbor's sole policeman, his face was drawn and miserable. Ben and Martin helped lift Connor's body into the dingy and then had to make their way back up the cliff. As they came to the station buildings and saw the students sitting about in small groups from which an occasional sob could be heard, Ben placed a hand on Martin's shoulder and gripped it firmly once more. "All hands on deck now," he said gently. Martin nodded, and they dispersed to console the students as best they could.

The Bugs were a sorry looking group when Reverend Nye arrived. He was a jovial retired minister in his seventies whom Trevor had asked to serve as unofficial chaplain to the station for the summer. He spent most of his time in the village, just coming to the station each Sunday to deliver a simple sermon. But in the face of suffering he took his job seriously enough. The professors seemed openly relieved when he appeared and took on some of their burden. Even Ben.

Louise told him what had happened, and because he had sometimes accompanied students on the cliff walks as well, Reverend Nye asked softly, "Who was on watch?"

Martin overhead the question and a wave of nausea came over him once more. Which meant his face betrayed a great deal when the minister looked him in the eye.

"You okay, my son?" He spoke with such tenderness that Martin could have cried. "This was gravity. And a foot placed wrong. Nothing more."

Martin nodded. "I know."

"Yes. Still. One can know something, and not really believe it."

He placed a hand on Martin's shoulder, and for an instant Martin thought the man was going to say a prayer over him. But the minister just left his hand there a moment and then moved on to the students.

It was still dark the next morning when Louise came to Martin's tent carrying a lantern. He'd been warned of her approach by the light dancing across the tent's previously pitch-black canvas. Martin stood up from where he lay in his cot, wide awake. The light that fell over her face as he opened the tent flap betrayed evidence of a sleepless night.

"Could you go back to Seattle with the professor, Martin?" She still called her husband that. Combined with her gentle teasing of Trevor for his absent-mindedness, it always sounded like an endearment. "He's going to need all his strength to confront the boy's parents. I would go, but I'm afraid the girls bothering him on the journey home with their little tantrums and cares won't be very helpful. The more hours and space he can put between himself and fatherhood the better."

Martin shook his head. "You don't think that's possible."

"No," she whispered with a slight smile. "But I still think you'll be a better companion for him. See if you can convince him to sleep on the journey. Pete Harrison's asked to go too. They were good friends, I think."

So Trevor, Martin, and Pete walked the mile to the ferry dock before the rest of the students awoke. The salmon cannery on the southern wharf lay dormant until the fish runs began in late summer, so the village was almost silent as they embarked for the boat ride to Seattle.

Usually the journey home passed quickly for naturalists on the lookout for eagles, whales, seals, and otters. Rocky islands rose from the dark green water at steep angles before rounding into gentle slopes. Fir trees and wiry red madrona trees gripped the gray rocks, which in turn were covered by golden brown grasses and mosses. But Martin didn't notice much of anything now. Eventually he was able to convince Trevor to lay down on one of the benches.

Pete stared out the window. Martin's first instinct was to let him be and walk the decks. But there was something so downcast in the bent of the lad's shoulders that Martin stayed where he was, in a seat across from him. They sat in silence, watching a small group of Lummi Indians paddle toward Lopez Island. A few minutes later, when they could no longer see the canoers, Pete noticed Martin had closed his eyes.

"You should take your own advice and lie down," Pete said.

Martin looked up.

"I'll be home soon. Trevor won't be."

They fell silent again, until Martin added, "It's good of you to do this, Pete."

Pete said nothing for a long moment, then he spoke.

"What do you think he'll say to them?"

Martin shook his head. "I don't know."

"What would you say?"

"I don't know that either, Pete."

Pete stared out at the water.

"I suppose Professor Cardiff would say they best buck up and face facts," he said. "Anyone tempted to see God's mysterious hand in Connor's fate is just taking refuge in a fable from some lower stage of civilization when men and women's fears of death drove them into a world of myths and fairy tales."

"He wouldn't say that aloud, Pete. Not to Connor's parents."

Pete made no reply and instead looked out at the water again for a long time.

"We had a lot in common," he said finally. "Connor's dad's a minister too."

Martin turned away from the passing islands. "I didn't know your father's a minister."

"Pretty orthodox, too." Pete replied. "We both came here opposed by our families. Guess that's something that drew us together. Seattle's a sin-soaked den of iniquity in their eyes. Especially the university. And the libraries."

"The libraries?"

"Yep. I can't blame them, really. When I came to college I believed as I had been taught. There's a heaven for good people, and a hell for bad people. I believed in a God who created Adam and Eve. Now I know that man evolved from protozoa, and that the Bible is great literature but a poor guide to the origin of man and his history. So, yes . . . the libraries."

Martin looked out the window again. Pete's words explained a great deal, about what he noticed in poems about ichneumon wasps.

"Does Trevor know this? About Connor's parents?"

"Yes. I told him. That's why I offered to come. Not sure how they'll take an atheist scientist breaking the news to them."

"Trevor's no atheist," Martin said.

"My dad would say he is. Or might as well be."

"But he believes in a God. Or something like it."

Pete shook his head.

"Not good enough. Dad knows some men who believe that a deity who works via natural law is a grander Creator than one who works by miracle. Who believe that man is not degraded, but ennobled, by having arisen from such a process, and that a God who creates through natural law is closer to, not farther from, man. But he says that's pantheism. Not Christianity. He says any evolutionist who pretends to be a Christian is just protecting the salaries they draw from institutions supported by devout men and women. When in actual fact they are undermining the foundations of faith."*

Martin's brow furrowed. This was a hard way of speaking of some of his beloved colleagues. He'd known plenty of museum entomologists back east who believed in evolution and still called themselves Christian. It was a long tradition among naturalists at the museum in Harvard.

"I don't blame him, really," Pete continued. "Most of these evolutionary versions demote Jesus to a good man and a great teacher. Because anything more would depend upon a miracle. Doesn't quite have the same punch, does it? No promise of an afterlife. Tells a man to be moral solely because ethics has by experience been good for the race. I guarantee you the knowledge that I'm being taught by men who believe in that kind of God is of little comfort to them."*

Martin did not know what to say, so he waited in silence as Pete looked out the window for a long moment.

"My parents believe everything happens for a reason. That God's will is in the fall of every sparrow, and there's a purpose to all suffering. That what we call evil will one day make sense. In heaven." He took a deep breath. "My sister died last summer, professor. Doctors couldn't get her baby out. Minister on duty said the fact mothers die in childbirth is evidence of the justice of Christ's atonement for our sins. Because the fact women so willingly risk death to attain motherhood shows that sacrifice is a law of life. Oh, and get this: women's sacrifices demonstrate God's handiwork in their character." Tears appeared in his eyes as he spoke. "But Connor wasn't sacrificing anything. He just put one foot in the wrong place and slipped. Where's the meaning in that?"*

Martin felt the grief and anger in Pete's voice too keenly to reply. He looked at the floor. Had he or Ben checked everyone's shoes? Had Connor been wearing boots, which might grip the rocks better, or sneakers? All those hours beside him on the rocks, and he couldn't remember. A line from Robert Ingersoll's lectures suddenly passed through his mind, one his father found particular comfort in, and repeated often, so that Martin knew it by heart:

Nature, so far as we can discern, without passion and without intention, forms, transforms, and retransforms forever. She neither weeps nor rejoices. She produces man without purpose, and obliterates him without regret. She knows no distinction between the beneficial and the hurtful. Poison and nutrition, pain and joy, life and death, smiles and tears are alike to her. She is neither merciful nor cruel. She cannot be flattered by worship nor melted by tears. She does not know even the attitude of prayer.*

Martin suspected Trevor believed this. But it was not something one could say to a mother and father who had lost their son. Or a young man who had just lost a good friend.

In the space of Martin's silence Pete bent down and rummaged through his satchel. He pulled a book out and handed it to Martin. "Bryan says something similar in this book. Dad sent it to me."

Martin opened the brick-red book to its title page.

In His Image

By William Jennings Bryan

"So God created man in his own image,
in the image of God created he him."
—Genesis 1:27.

Martin turned the book over in his hands. "Why didn't you mentioned these pulls on your conscience before?"

"Most of the time I try to forget them. Isn't easy, since I live at home still. But when I'm out here, and working through a specimen drawer, every single one of my parents' fears fade away. I simply can't make sense of these creatures without assuming forms have descended from other forms. The useless parts, the groupings, the geographical patterns. Then the mysteries and puzzles become overwhelming. I must try and figure it all out, and pick up the next tray of specimens."

Martin smiled. This was why he and Pete got along so well.

"Of course they see only worldly temptations in those trays," Pete continued. "And sinful pride that I might understand any of it. The beetle specimens might as well be the women and wine of olden times."

Martin handed the book back.

"Could you do me a favor, professor?" Pete said, not taking it. "Tell me what to say to him. My dad."

"Oh, Pete," Martin protested. "I can't tell you what to say to your father about Bryan. One can't get these things secondhand like that."

"Bryan has some pretty hard things to say about you," Pete pressed. "I mean college professors. Evolutionists. It's all made Dad pretty anxious. Angry even. But maybe you could talk to him? He'd find out for himself that you're not, in fact, a monster."

"Does Bryan say that?"

"No. Not quite his style. But one can find it between the lines easily enough." He paused, trying to find the words, or stop the tears. "Mom and Dad are afraid the only thing I'll be good for after college is eternal damnation. It's a hard thing to see my parents fear. I'm pretty sure when they hear about Connor they'll panic and ask me to quit college. He died before returning to the fold, you see. That's what they'll say." His voice trailed off and he took a deep breath. "But I think that if they could meet you. Talk to you. See that you're a good man."

"Let me take a look at what Bryan says," Martin offered, pained by the tone of Pete's voice and his dilemma. "But I'm afraid I can't promise to be much use."

"Thanks, professor. If you could just read the chapter on the origin of man."

"They don't all go together?"

"Yes. But isn't subjecting you to his arguments against Darwin enough to ask?"

Martin smiled slightly and nodded. "I suppose so."

⁓

The next day, as the afternoon sun warmed the science building in Seattle to an uncomfortable degree, Martin was in his office sorting through some specimens from the Experiment Station in Puyallup that had arrived while he was in the field when Trevor appeared and sat down. There was a weariness in his eyes, for Connor's parents had arrived from eastern Washington to retrieve their son's body. He idly picked up a tray of lady beetles and looked at them in silence for a moment.

"Beautiful little creatures," he said finally. He set them back down very gently. "I've never wanted to believe so strongly that these little species are each the special creations of a benevolent God. I'd have been more helpful."

"Pete said he told you about their beliefs."

Trevor nodded. "I'd not have known otherwise. I might have said something very hurtful."

"I doubt that very much," Martin countered.

Trevor sat back in the chair, his whole body betraying exhaustion.

"You ever heard Billy Sunday preach?"

"No," Martin replied.

"I heard him once. During the war. God, it was powerful. Maybe he knew a couple of us were in the audience. Demanded that scientists just try and take their explanations into a room where a mother has lost her child. He thought we'd tell her the child died because he wasn't fit to live. Swore that if our speeches about eugenics, evolution, and the fortuitous concurrence of atoms hadn't driven her mad, he'd take the Bible and read God's promises, and her soul would be flooded with calmness like a California sunset."*

"You wouldn't have said anything about the survival of the fittest even if they were atheist," Martin said in the ensuing silence.

"No." Trevor replied. "Guess I wouldn't have known what to say in that case either." He pressed his lips together tightly for a moment. "Having young Harrison there helped a great deal. He didn't say much either. Told them how proud he was to have their boy as a friend. How well liked he was by the students. That meant something." He paused again. "It was a courageous thing. To come back with me and face them. He didn't have to."

After Trevor left, Martin looked at the piles of specimen trays on the floor, tables, and cabinet shelves. Then he looked at the dozens of glass vials still in the box before him, each containing a small insect with a blank label on which he was supposed to add a name. A few hours of work, and the task would be done and he could send the identifications to the Experiment Station. Instead he packed up his satchel and walked home.

He went out onto the deck with Pete's copy of Bryan's *In His Image* in hand. Mrs. Macleod had all the windows on the first floor open and Al

Jolson's voice singing "Mandalay" drifted from the radio below. Since Pete had given him permission, Martin skipped past Bryan's opening chapters on the Bible to the chapter entitled "The Origin of Man" and read Bryan's passionate broadside against evolution. He didn't read things like this very often, unless at his dad's behest. But he'd been given a task, and sat with a notepad and pen at hand to try to fulfil it as best he could.*

Bryan spent the first several pages trying to ridicule evolutionists' story of man's origins into oblivion. He used Darwin's infamously cautious wording in *On the Origin of Species*, all the "probably" and "apparently" and "one can imagine" and "we may well suppose" (such phrases occurred eight hundred times in the book, Bryan wrote in amazement) as evidence that Darwin was guessing. Americans were being asked to give up belief in God for a guess.

Martin paused his reading and jotted down a few words about the role of hypotheses in science. He made a note of the rule that explanations in science must be naturalistic, rather than supernatural, lest there be no control on what a man might imagine. And he outlined an argument for why Darwin used such cautious, tentative words in attempting to explain a past he could not actually witness.

He heard the door to the apartment open, his father's footsteps, and something being dumped on the kitchen table. Will came out to the deck.

"I've been thinking, Marty," he said as soon as he sat down on the other deck chair. "I shouldn't have bitten your head off for pestering me."

Martin smiled. Eight weeks might have passed, but the minute they were in the same room together his dad charmed away any conflict between them.

"And I didn't mean to pester," he replied.

"But you're here because of me. I know that. No surprise you're worried your old man'll get you kicked further west. Not many places left that direction to go."

"That's not what I'm worried about, Dad."

"Okay. Well. That's what I'm going to tell myself you're worried about. Because I don't understand the other stuff."

"Fair enough."

Their truce struck, Will told him all about his summer on the Tulalip Reservation. The stories inevitably turned into a speech about everything that was wrong with the Bureau of Indian Affairs.

"Governor Stevens had the treaties all written out before he even arrived, Marty. Read 'em to the Indians through an interpreter, and got a couple of chiefs to sign under the convenient assumption that those men could represent the will of human beings from the Columbia to Canada. No need to ask questions about whether the assumption was right or the treaties just. Not when the progress of civilization is at stake. Treaties promised the tribes schools, Martin. That's one of the reasons they signed and went to the reservations. Now the men can't leave the reservation without the Indian agent's permission! Everyone's supposed to be a farmer now. Trouble is most of the reservation land isn't good for farming. Tulalip doesn't even have enough freshwater for everyone. Think the Indian chiefs understood that's what having a treaty with whites is all about? They're the real scraps of paper around here."*

Eventually, because Martin listened mainly in silence, Will asked about the summer at the biological station, adding, "I didn't expect you home yet. Thought it went to the end of July." So Martin explained why he had come home before the session ended.

"I'm very sorry, Marty. Was he a student of yours?"

"No." He took a deep breath. "Not officially. He worked in the collection room a lot. Plants mostly. He would have made a good taxonomist." He paused again, as his father waited. "Trevor had to speak to his parents. Cut him up pretty badly, I think."

"'Course it did. He's got a heart. And children. And imagination," Will said quietly.

"He didn't know what to say." Martin replied.

"What is there to say?" Will said with a shrug. "Probably a lot to be said for your silence. In cases like this."

He pressed Martin's shoulder gently as he spoke, and then went back inside. Martin had tucked Bryan's book under his elbow so Will could not see the title on the spine. He didn't feel like listening to his father's opinion on Bryan just now. Left alone, he opened the book again and read Bryan's description of a string of absurd scenarios concocted by confident evolutionists to explain how eyes, legs, and the big toe had arisen, peppering the account with mini-editorials: "Can you beat it?" and "It just happened so!" Bryan expressed astonishment that any person intelligent enough to teach school could talk such "tommyrot" to students and look serious while doing so. Having read *The Descent of Man*, he ridiculed Darwin for attributing the evolution of the powerful male mind to a struggle for females while blaming the evolution of male hairlessness on female preference. "Unless the brute females were very different from the females as we know them," Bryan quipped, "they would not have agreed in taste. And how could the males have strengthened their minds by fighting for the females, if, at the same time, the females were breeding the hair off by selecting the males? Or, did the males select for three years and then allow the females to do the selecting during leap year?"

Martin had never read *The Descent of Man*. Will wanted to see how a man's ideas had been put to use in the world, which meant his library was full of books written by Darwin's successors, rather than Darwin's books. By the time Martin became an entomologist with strict rules about focusing on nonhuman animals, *The Descent* had become irrelevant to his own work. Now, reading Bryan's account of Darwin's attempt to apply his theory to human beings, Martin was not surprised that Darwin was carried away by some of his examples. Darwin had tried to transform how his fellows explained human origins. Surely the first pass at such an extraordinary revolution would be imperfect. But Martin also knew the careless enthusiasm with which some of his own colleagues imagined how complex traits evolved. He started making a list of more careful authors and more convincing examples. He could arm Pete with better explanations of adaptations, while drawing attention to rudimentary structures, geographical distribution, and extinction as phenomena that evolution explained better than the theory that God created species as they are.

But then Bryan moved to what he called the "natural tendencies of Darwinism." "A groundless hypothesis—even an absurd one," Bryan wrote, "would be unworthy of notice if it did no harm." Yet evolution, he continued, did do incalculable harm, for the natural and logical effect of the theory on young Christians was to lead them away from belief in a personal, loving God. Martin's pen and notepad soon lay forgotten, as he read Bryan's detailed argument that in undermining faith in God, the doctrine of evolution destroyed belief in the brotherhood of man, moral responsibility, and any confidence in the ability of a man to avoid temptation. The destruction of belief, Bryan argued, had profound, terrible consequences. Recent history—the survival-of-the-fittest doctrine of industrialists, the war, the movement to scientifically breed a higher race of human beings—were proof enough, in Bryan's eyes, that humanity was incapable of true goodness once men and women fell adrift from faith. Once men accepted Darwinism, Bryan warned, "there is no act of disinterested love and tenderness, no deed of self-sacrifice and mercy, no aspiration after beauty and excellence, for which a single reason can be adduced in logic." *This*, Bryan insisted, was why he rejected the Darwinian doctrine.

Bryan made much of how Darwin had lamented the fact that, while in savage societies the weak in body and mind were soon eliminated, civilized men checked the progress of elimination through poor laws, medicine, vaccination, and asylums for the imbecile, the maimed, and the sick, thus preserving thousands with weak constitutions. Darwin's words—"No one who has attended to the breeding of domestic animals will doubt that this must be highly injurious to the race of man"—were proof enough, Bryan wrote, that his theory put man on a brute basis and attacked the very foundation of Christianity. They were proof enough, that once man believed he had reached his present perfection by a cruel law under which the strong kill off the weak, then, if there is any logic that can bind the human mind, we must turn backward toward the brute if we dare to substitute the law of love for the law of hate.

Bryan conceded that in the end Darwin's own heart rebelled from the "hard reason" on which his heartless hypothesis was built. For Darwin had, in the end, insisted that the instinct of sympathy that had led to charitable institutions and hospitals was far more important than the prospect of improving the race. "But some of his followers," Bryan warned, "are more hardened." He cited as evidence a scientist's argument against prohibition on the grounds alcohol rendered a service to society by killing off the degenerates. "Can such a barbarous doctrine," Bryan demanded, "be sound?"

Pete was right. Professors of biology and zoology came out looking pretty poorly by the time Bryan was done. With trusted access to the minds of millions of Americans, Bryan attributed the terrible temptations to reengineer man and society on biological grounds, whether through eugenics or industrial or national competition, to the inevitable result of the biology professor giving his student "a new family tree millions of years long" and then setting him adrift, "with infinite capacity for good or evil but with no light to guide him, no compass to direct him and no chart of the sea of life!"

After finishing the chapter, Martin set the book down and looked at his notepad, blank since he had stopped scribbling a list of more careful explanations of adaptations. They didn't seem very relevant by the time Bryan was done explaining why the moral compass and salvation of beloved sons and daughters were at stake.

Chapter 6

When Pete arrived in his office a few weeks later with a tall gentleman by his side, Martin knew the guest must be his dad before Pete said a word. The resemblance was striking, as though Pete stood before him, just a few decades older. Reverend Harrison shook Martin's hand with a firm grip and, after insisting that Pete show his mother, who was waiting outside, the campus while he and the professor got acquainted, sat down with the manner of a man who was going to share the latest gossip rather than accuse Martin of destroying his son's chances of everlasting life. He began by saying that Pete spoke very highly of his courses, though Martin wasn't sure his guest meant that fact as a compliment.

"There's much novelty in it, I assume," he added.

Martin shook his head. "Most of the time they're bent over insects that naturalists cataloged long ago."

Reverend Harrison looked closely at a tray of specimens on Martin's desk.

"Yes, but you explain these little creatures as the result of evolution over millions of years," he said. Then he looked up. "Peter said he told you how he was raised. Of our beliefs. It is a novelty to him, professor." He paused, and then pulled a folded newspaper from his bag. "I brought something rather specific to ask you about, if I may."

Martin took the paper and read the headline.

Geneticist William Bateson Holds that Darwin Must Be Abandoned. Theory of Darwin Still Remains Unproved and Missing Link Between Monkey and Man Has Not Yet Been Discovered by Science. Claims Science Has Outgrown Theory of Origin of Species. Distinguished Biologist from Britain Delivers Outstanding Address on Failure of Science to Support Theory That Man Arrived on Earth Through Process of Natural Selection and Evolution of Species. Have Traced Man Far Back but Still He Remains Man.

The missing link is still missing, and the Darwinian theory of the origin of species is not proved. While declaring that his faith in evolution was unshaken, he frankly admitted that he was "agnostic as to the actual mode and process of evolution." Believing in evolution in "dim outline," he pronounced the cause of origin of species as utterly mysterious.[*]

Martin had heard about this speech. Ben called it the Sarajevo shot that had launched Bryan's campaign, for Bryan had decided that if biologists couldn't agree on how evolution happened then it certainly shouldn't be taught to America's children, not when it had inspired a war. Martin had noticed Bryan included a footnote to Bateson's speech at the end of *In His Image*, as evidence of evolutionists' uncertainty.[*]

"For what it's worth," Martin said, handing the newspaper clipping back, "many biologists found Bateson's statements quite misleading."[*]

"A bit nervous the myth of consensus was burst?"

Martin took a deep breath and was silent for a long moment, but Reverend Harrison must have been primed by Pete to let him have his pause.

"Perhaps," he said finally. "But Bateson wasn't saying anything new. Darwin never claimed he'd proved selection can create new species. Even today, few naturalists think natural selection explains all. Some insist it explains nothing. But take any particular pattern explained by evolution, and the claim that evolution explains facts better than the theory of special creation still stands, whether natural selection is the mechanism of change or not."[*]

"Some believe quite the contrary: that evolution does not in fact explain the most important facts of our existence. Man, to take just one random example."*

Reverend Harrison winked as he spoke, and Martin smiled. At least his guest had a sense of humor, despite the stakes.

"I'm not sure I've observed man enough," he replied. "I study the boundaries between species of insects."

"Well, then. I'm listening. Why do they convince you?"

Martin picked up a tray of Anthribidae he had been working on and placed it in front of the minister.

"This is all one species of weevil. See how much variation there is? When taxonomists worked under the assumption that species are independent creations of God they still explained variation like this—a darker color here, a slightly larger tail there—as the influence of the environment. They saw all of these varieties as united by common descent from an original species created by God. But even the best naturalists couldn't always decide whether some specimens represented a new species or simply a variety. Darwin argued that surely the distinct modes of creation—one by God, the other by the environment—would leave some mark. It doesn't. But under the theory of evolution, if species are in fact only well-defined varieties, the difficulty of distinguishing species from varieties becomes expected. And that's true no matter how that divergence came about."

"But we're not all taxonomists, are we?" Reverend Harrison pressed. "We're not all gripped by such technical puzzles. If this is what convinced you, why claim, as Professor Osborn does, that there is a flood of proof of the evolution of man and any opposition is irrational?"

"Some don't," Martin replied. "Much of my own work proceeds the same whether one sees species as created or not. But when I sit back and think about the patterns before me, about why a distribution falls the way it does, God made it so doesn't satisfy."

"But why not see that order as reflective of God's plan? Why press God and faith out?"

Martin leaned forward, trying to engage in earnest, for Pete's sake.

"But there is a tenet of faith in science. That nature's laws are invariable. That there is constancy of cause and effect. That under the same circumstances the same thing will happen always and everywhere. Without this assumption science halts."*

"So you admit your belief in the constancy of natural law is based on faith," Reverend Harrison said, "when in fact you have no proof of that constancy. When surely a God who can establish natural laws can suspend them when it suits his purposes."

Martin was out of his element and he knew it. He crossed his arms across his chest to ward off the adrenaline rising in his blood as the minister continued.

"And if science is knowledge gained and verified by exact observation, what exact observation do you have of the spontaneous generation of life? Of one species changing into another?"*

"I'm a naturalist, sir," Martin replied. "I'd be the last one to speak against science as exact observation or even against the constancy of species. Within a certain time frame. All my work depends on both. But relying strictly upon observations of present-day organisms will tell us only so much about the past."

Reverend Harrison shook his head. "On the contrary. I believe that careful, pious attention to present creatures will convince an open-minded, truth-loving man that it is impossible to deny that we and all creatures on earth were created by a wise and good Creator. The God of the New Testament. Have you ever watched a great blue heron, professor? Really watched one? You can see how his legs are formed for wading. You can observe how nicely he folds his feet when putting them in or drawing them out of the water. He does not cause the slightest ripple, and the fish has no signal of his approach. His entire frame is beautifully designed for the task at hand. We have no experience with such extraordinary design separate from the existence of a designing mind."

Martin said nothing. Chasing down descriptions of Anthribidae from long ago, he had spent part of his working life immersed in careful accounts of insect anatomy and behavior accompanied by reverent asides that such purposeful design reflected the wisdom, power, and goodness of God. It was a tempting explanation. The adaptations were extraordinary. The diversity awe-inspiring. But even before his mother's death, he had been inclined to sympathize with Robert Ingersoll, who demanded of such arguments as Reverend Harrison gave now: Yes, but isn't the arrangement a little tough on the fish?*

His guest switched tactics, given Martin remained silent.

"And the fact scripture says that God created every living creature according to its kind?"

Martin shook his head. "I'm afraid we disagree very much on the kinds of evidence allowed to understand the past."

"Because you don't allow Scripture."

"No," Martin replied. "I believe it as poor a guide to the origin of species as it is to the question of whether the earth goes around the sun."

Harrison slowly smiled. "Come now, professor, stay above the belt. There's a great deal more at stake in a theory that contradicts the biblical account of man's origin, than one that contradicts poetic descriptions of the shape of the universe."

"But if you admit poetry in one case, why not the other?"

"Try to provide a poetic interpretation of 'So God created man in His own image' if men came from apes. I don't understand how a thinking and feeling human being can find the effort anything but revolting."

Martin fell silent again, as he tried to understand what possible relevance revulsion could have in deciding a scientific question.

"Now, professor," his guest said with a gentler tone. "What I want to know is, do you tell them of the alternative? Do you tell them of the belief that they were created in God's image, with a divinely implanted rational and immortal soul that can discern good from evil? Do you tell them of the theory that through God's grace they can be saved?"

"No," Martin answered firmly. "I'm sure you would agree that under the circumstances I should leave that alternative to you."

"We have a tough job of it after you've told them they're all apes, though, don't we?"

"Most of the time I'm just trying to help them understand a bunch of insects. But even if I did talk about human evolution, or lecture them on the evidence for humanity's common ancestry with apes, that's a very different thing than telling them we're all apes. One might still conclude we are human beings. Fully and completely."

Reverend Harrison shook his head. "No. You need not mention man at all, if you like, and it all amounts to the same thing. Materialism. Bestialization. Through and through. It is dishonest to argue otherwise."

Martin frowned, but he replied firmly, despite the trembling he felt setting in.

"Surely one must see as much earnestness in another man's efforts as one's own, and acknowledge that for all men it is a difficult thing to determine whether one's ideas of God, or of nature, are true."

"But that's just it, professor. It is not a hard thing to determine, if one accepts certain doctrines as fundamental." He counted them off on his fingers with reverence in his gaze, as if the items were there to be seen. "Belief in the Bible as the inspired and final authority in life; belief that Jesus Christ was begotten by the Holy Spirit, born of the Virgin Mary, and is true God and true Man; belief that man was created in the image of God, that he sinned and thereby incurred physical death and spiritual death; belief that the Lord Jesus Christ died for our sins, and that all who believe in him are saved; belief in the resurrection of the crucified body of our Lord and his ascension into heaven; belief in the bodily resurrection of the just and the unjust, the everlasting felicity of the saved, and the everlasting conscious suffering of the lost."*

It was hard to speak within the silence that engulfed the minister's final words, but Martin forced a reply.

"That is a demanding list."

"And absolutely irreconcilable with a materialistic explanation of man's origins. Biologists replace the fall of man from a state of innocence with belief in the ascent of man from a primitive condition. They insist sin is only the remains of the ape and the tiger in us, and so what then, remains of guilt? Of conscience? What guide is there for man, left in the terrible, meaningless, darkness of his sins? What meaning is there in the pain and suffering that comes his way? How will he know right from wrong, and why should he not behave like the beasts, motivated solely by his greed and his belly?"

This time Martin did not pause before replying.

"No. By your account of things, I should be on all fours fighting you with my teeth and claws rather than my intellect. But in fact we both can sit across from each other and speak with good will, despite our disagreement. I don't believe in scripture, yet I believe in humanity and brotherhood rather than war."

Reverend Harrison met Martin's words with a steady gaze and firm voice.

"The individual agnostic of virtue is not enough to counter the last decade's evidence, professor. One must judge an idea by its fruits. Under Christianity we had an increasing belief in universal brotherhood, peace, and the Golden Rule. But under the spell of Darwin's godless universe and Nietzsche's superman, Germany claimed the right, the natural right, mind, to pursue the struggle for existence in earnest and destroy weaker nations. The Germans declared their scientific theory of man and the fundamentals of their faith and enforced their biological doctrine with shells and mustard gas. And if men believe man is an animal whose instincts commit him to selfishness, violence, and war, then how does one prove the militarists wrong? Once men are looked upon as children of apes, what does it matter whether they are slaughtered or not? When science proves that might is right?"*

"Science tells us nothing of rights," Martin countered.

"But scientists have," Harrison replied. "And not just in Germany. Isn't this trial in Chicago evidence enough? Darwin's theory destroyed those boys' sense of morality. And now they're being defended on the grounds they have no free will and are biological machines who cannot be held responsible for murder."

Martin cringed within at the stakes being raised so high. Since he'd returned home Will had insisted on reading news clippings about the trial of two adolescents who had confessed to kidnapping and murdering a schoolmate in Chicago. One of Will's heroes, the defense lawyer Clarence Darrow, had signed on to the case in order to prevent them from being hanged. Will had read Darrow's closing statement aloud, so Martin knew that Darrow argued against capital punishment on the grounds the defendants were diseased machines, entirely determined by their heredity and environment. Darrow had called for pity for his "boy clients" rather than revenge, because it wasn't their fault they did not know the difference between right and wrong.*

"Did you know one of those boys told a reporter that a thirst for knowledge is commendable, no matter what pain or injury it may inflict upon others?" Reverend Harrison continued. "And that a boy is justified in pulling the wings from a fly if by doing so he learns that without wings the fly is helpless? He got that from a biology teacher in Chicago."

"I doubt that very much," Martin replied firmly. He tried to switch targets. "Surely men have manipulated the alternative just as easily. To match their own desires."*

Harrison shook his head. "No. Not just as easily. Scripture demands do unto others as you would have done unto you."

"Scripture demands, but how often have men obeyed? Men have slaughtered each other in the name of Christianity as well. Turn the enemy into infidels, as we did to the Germans, and men can imagine God approves of our blowing them to smithereens." He paused. The trembling would be difficult to hide if they went on much longer. He tried to speak with a tone of finality. "But you don't believe the things fallible priests and parsons have done in Christ's name undermine Christ's demand that I love my neighbor."

"Do you think it so easy to separate what men have done with Darwin's ideas from the belief man evolved from the apes? And that you can find anything good and virtuous in the belief that we arose from the violence and strife of an interminable struggle for existence?"

"But Darwin didn't invent those things. He didn't invent the violence and suffering in nature. He observed them all around him. In his own society. And he tried to explain them."

"We have an explanation, professor. In man's disobedience against God and the depravity of original sin."

Martin had no answer. He was being asked to debate something on which he had no firm convictions except that to speak with confidence on any of it made no sense, and his whole frame seemed to rebel from the effort. He had staved off the trembling but had broken out into a cool sweat as he concentrated on forming coherent replies. It was the clock striking the lunch hour that saved him, for at that signal Reverend Harrison took his hat in hand and stood. Martin wiped his hands on his trousers before standing. They had not spoken of Pete. Not directly. Then Reverend Harrison asked one final question as he stood:

"What, professor, of the mother who spends eighteen years preparing her boy to go out into the world a believer, only to have that education smashed in fifteen minutes by professors who tell him his Bible is a fable and he was not, in fact, made in the image of God?"

Martin said nothing, and Reverend Harrison extended his hand.

"With all due respect, Professor Sullivan, to hell with your science if it's going to damn souls."

"We can at least agree on the hope that it will do no such thing," Martin replied, trying to press a signal of his sincerity into the minister's palm.

⁓

When he saw Pete again a few days later, his young friend made no reference to his father's visit. But he seemed distracted. Martin had asked him to help with this collection room as well, on the grounds he had done such good work at the station. Martin kept the second motivation, that he wanted to keep an eye on him and keep him busy after so tragic a summer, to himself. Pete had worked on recording a stack of specimen trays for an hour before he suddenly swore beneath his breath and tossed his notepad on the floor. The sound caused Martin to look up from his own set of specimen trays.

"I'm sorry, professor," he said. "My heart's not in it today. Or my mind. Whatever."

Martin feared his conversation with Pete's dad had only heightened the family's resolve to have him leave the university. But it became evident that Pete wasn't thinking about his dad when he asked whether Martin knew Phoebe was leaving.

"She's taken that position as an assistant in Morgan's fly lab. In New York."

Martin added his observation of Pete's distress at this news to the pattern of Phoebe's visits to the collection room since everyone had returned from the station. Thanks to Ben's teasing, he had paid more attention, and noticed much more in their interactions with each other than he had before. He now wondered whether the real target of Reverend Harrison's visit to campus had been to meet Phoebe. No doubt her unorthodoxy was damning evidence of everything wrong with the world. Rumor had it that the dean of students had recently reprimanded her for attending a meeting of the Seattle Birth Control League. That linked her, by association, with some of the most radical reformers in the city. Phoebe was not the woman to bring Pete back to the faith.

"You know how good she is in the field," Pete was saying. "And now they'll put her in front of some milk bottles and a microscope and have her counting hairs on flies. A month and she'll be giving speeches about how work must be experimental to count as science."

Martin shook his head. "I'd give Phoebe more credit than that, Pete. She's already one of the best experimental thinkers among you. I don't mean the kind that sits at a lab bench and insists on messing up nature to figure out how it works. The naturalist who compares one fact to the next to test their ideas counts just as much. And she's better than any of you in the field. One doesn't lose that."

Pete had finished packing his satchel, and stood. "Of course you're right," he nodded. He went to the door, then paused. "By the way, I never thanked you for talking to Dad."

"You're welcome. I'm not sure it did much good."

"I think it did," Pete said. "There aren't many men on the other side who will give him the time of day. I think you astonished him a bit by not lecturing him about dog breeds and tail bones." He paused. "You know, he can actually follow our thinking pretty well up to a certain point. But says too much mental gymnastics is required to believe both that man arose from apes and that he's anything very special in God's sight. I guess we aren't all skilled in acrobatics, mental or otherwise."

"Oh, I'd say we're quite good at it," Martin replied. "Every one of us. That's the whole trouble."

"Honestly, professor," Pete said with a smile. "You're enough to make a man give it all up and go fishing."

"How about just doing more science. And better," Martin answered, returning the smile.

Martin thought they were done, but Pete hesitated at the door.

"I probably won't be here next term. Dad's got me a job teaching a nature study class for the primary school run by his church. So that's something."

Martin looked up again. He wanted to speak a dozen arguments why Pete should disobey his parents' command. Instead he asked, "And when a kid asks you why male birds are so brightly colored?"

Pete shrugged and smiled again. "I'll ask them, who can say?"

As he turned to go he had to stop abruptly to avoid crashing into Trevor, who suddenly appeared in the doorway breathing heavily. Pete's footsteps disappeared down the hall as Trevor collapsed into a chair.

"Harrison okay?" Trevor managed to get out. "Been worried about him."

"You aren't the only one. His parents want him to quit university."

Trevor nodded and took a deep breath. "I was afraid of that. Seems we're losing two of our best students. Did you hear Miss Bartlett's taken a position in Morgan's lab?"

Martin nodded toward where Pete had been standing. "He's a bit cut up about that too."

A long "ah" escaped as Trevor put two and two together.

"They're very young," he said. "Can't tell 'em things'll be fine in the end. Doesn't help to say it's only a year, since she's not coming home to-morrow. One can't really reason with a man on such matters."

Having caught his breath, he leaned forward to signal a change of subject.

"Well. I've got some news, Sullivan. Brace yourself. Old Weyerhaeuser's agreed to pay for a building for a natural history museum."

Martin stared at Trevor's giant grin. "Are you kidding?"

"Seems my dressing up in black tie for that gala and catching him for five minutes did the trick. Louise was right, as usual. Now. I want you to go down to Tacoma and look at a collection of Coccinellidae on offer from an old minister. Name's Reverend Gray. I've been putting him off for years because we didn't have any space, and he might have changed his mind. If it's a good collection, we should get it."

Martin didn't feel like spending time with another minister but nodded. Then he recovered enough from his astonishment to ask if there was money for a full-time assistant curator.

"I don't see why not. They seemed inclined to trust our judgment for once, and I'll certainly put my hand to making the case. Harrison, right?"

Martin nodded.

Trevor had turned toward the door when he stopped. "Say. Since you'll be going to Tacoma, mind adding another errand to the trip?"

"Sure."

"Not sure you'll like this one, actually," Trevor replied. "You know James Slater, the biology professor at the College of Puget Sound?"

"Herpetologist?"

Trevor nodded. "He's looking for someone to teach a course on eugenics next summer. Evening class. I know the applied line isn't your preference, but mind looking him up while you're down there?"

"I don't know anything about eugenics," Martin replied.

He thought that was a fair enough statement, despite the fact he'd occasionally received an earful from his father on the topic. Soon after he'd announced he had accepted a teaching position, Will had taken it upon himself to prepare Martin for the prospect of being faced with a bunch of undergraduates eager to know what biology said about human problems. A few days before their departure for Seattle, Will had tossed a copy of the *Century Magazine* at him with the demand: "What is it with you lot, the fumes from all those preservatives?"

Martin had smiled and said he didn't think naturalists had that excuse anymore. His smile disappeared as he read "The New Decalogue of Science: An Open Letter from the Biologist to the Statesman" by a man named Albert Wiggam. He frowned when he came to Wiggam's claim that it was the "duty and privilege of biologists" to deliver biology's stern warnings of racial degeneration if the unfit were not prevented from mating.*

Martin turned a few more pages, clearly skimming. Will leaned toward him.

"The wowser insists improving the environment of the poor will do nothing to aid them because heredity is the only thing that determines the character and worth of a man. Says the claim all men are born equal is based on a great sentimental nebulosity and social classes are ordained by nature not economics. Apparently all the medical and technological progress provided by science will be a biological joyride with hell at the next turn. Isn't that fine? And the only way to create a better world is by choosing our mates scientifically, because biological laws prove that only heredity matters."

"I think we're pretty ignorant of biological laws," Martin replied, handing the magazine back.

"Yes, but look here, Martin," his father pressed, flipping to the end of the article. "Look at this list of biologists who agree with him. One of them calls it the most important book on the social significance of biology. Professor Davenport thinks it should be distributed in pamphlet form to every man and woman in America!"

Martin took the magazine in hand again and read Princeton biologist Raymond Pearl's declaration that Wiggam's biology was "absolutely sound so far as we know at present" and John Hopkins University biologist Edwin Conklin's words that the book provided a concise and trenchant statement of the "great teachings of science." The Clark University psychologist G. Stanley Hall had "read nothing anywhere since the war that seemed to me so sane and timely." And the editor of the *Journal of Heredity* insisted that "all biologists will agree that a copy of the article should be put into the hands of every public official in the realms of statecraft and national education."*

Martin had never paid much mind to this kind of work, though he knew many of his colleagues saw eugenics as the primary means through which biologists could contribute to social welfare. In 1921 in New York, during breaks from working through the collection of Anthribidae at the American Museum of Natural History, he had walked the exhibits of the Second International Congress of Eugenics, set up on the museum's ground floor. He knew the director of the museum, Professor Osborn, insisted that the museum must teach Americans that the laws of nature proved that biologically informed mate choice was a patriotic duty that would save the nation from racial degeneration. But he had always been bewildered by Osborn's confidence that human beings could be so easily classified, a basic prerequisite of the decisions required. In his more flippant moods it had all made him smile, given the hours and hours he spent observing thousands of specimens so he could say something certain about a few species of beetles of interest to no one.*

Martin had shrugged and passed the magazine back.

"I can't explain their enthusiasm, Dad. Isn't my line of work."

Will pulled out another clipping.

"Here's Wiggam again. Get this. Eugenics is simply the projection of the Golden Rule down the stream of protoplasm and Jesus Christ himself would approve. Doesn't that sound fine? He says we have enough science at hand to bring the world into an earthly paradise. All that's required is the courage to apply science to mate choice."

"Sounds very fine indeed," Martin said. He had returned to his own entomological journal and did not look up as he spoke. "But then, paradise generally does."

"Well. I don't understand how they can possibly think heredity matters a whit," Will pressed, "when the environment of so many is so poor."

When Martin made no answer, Will paused a moment and then

added his inevitable "on the other hand . . ." Martin could never understand how this habit did not immobilize his dad entirely.

"And yet. If criminality is due to hereditary feeblemindedness, then we can no longer see delinquents as sinners who must be punished. They are simply broken machines who deserve our sympathy rather than condemnation. And maybe there's something to be said for all this nonsense, if it makes men more understanding of their fellows. Perhaps even care for them better."

"Strange idea of understanding and care," Martin reminded him, "when it might include sterilization against one's will."

"Exactly. Exactly." Will nodded. "Still. I can see the logic, if a man's willing to give up individual liberty."

That was a big "if" for his dad, who didn't believe a man in his right mind should be willing to say of another man "I know what you are and what is best for you," whether civilization fell as a result or no. The Volstead Act drove him crazy. Though not much of a drinker himself (he loved the Industrial Workers of the World's warning that a man couldn't fight booze and the boss at the same time), he hated the man, whether scientist or minister, who would judge his fellow for giving in. This was clearly not the kind of science Will had in mind when raising Martin on a bunch of freethinkers who believed in the power of reason to make the world a better place.

With this conversation with his dad in his memory Martin was about to say no when Trevor added, "Best way to learn about a subject is to teach it to a bunch of undergraduates. Or at least consider teaching it."

Martin hesitated. He thought of how Bryan had cited Darwin's words from *The Descent of Man* as evidence that eugenic thinking was a terrible, natural tendency of Darwinism. And he thought of the fact that Trevor, whom he respected a great deal, had included a unit on eugenics in his class on evolution for more than a decade, and had just recently developed an entire course on the topic. Martin suspected he would eventually be asked to take his turn at the new course, billed as teaching "the principles of evolution in their relation to human welfare." They were a small department, and it was becoming one of their most popular courses since it was an elective for the library school and home economics students as well. And then there was the lifetime of his father's speeches about joining a cause to help mankind. Those memories meant he could not entirely ignore the mystery of why so many of his colleagues thought they'd found one in eugenics. Besides, he couldn't easily decline the suggestion that he learn more about a subject before deciding. So he took Slater's address in hand as well.*

After Trevor left, Martin scribbled on a notepad, *Don't decide yet. Possibility of position here. Martin.* Leaving the note on top of Pete's next stack of specimen trays, he felt a twinge of guilt as he did so. The note was a direct assault on Reverend Harrison's plans and hopes for his son. Then he told himself that if he was right about Pete's relationship to Phoebe, a museum job was the least of the minister's worries in the effort to bring his son back into the fold, and at least a museum job involved documenting the wonders of creation in some sense.

Chapter 7

A month passed before Martin was able to fulfill his promise to visit Tacoma. He'd received dozens of specimens from the agricultural school in Pullman with requests that he identify the little beasts as soon as possible. Finally, in late September, a brief lull in the rate of arriving packages and the fact students had an exam on Monday meant he could leave campus early on a Friday afternoon and make the trip to Seattle's scrappy, industrial rival to the south.

He decided to travel the five miles downtown to Colman Dock on foot. So he walked south to the center of campus, along a long pathway that passed the fountain built for the Alaska-Yukon-Pacific Exposition of 1909. Beyond, huge fir trees lined the campus at the edge of Portage Bay. He crossed over the Montlake Cut shipping canal that linked Lake Washington and Lake Union via a makeshift pedestrian bridge made from a series of barges tied together. Then acres and acres of undeveloped land, clear-cut in the last decade or so, and already earmarked by enterprising citizens for a future golf course for the rich and an arboretum for everyone else. Seattleites were intent, it seemed, on taming the landscape into a replica of the Scottish Highlands. There were fewer trees as he made his way downtown, past new construction and the remnants of a massive re-grading of the surrounding hills. It formed an imposing skyline for a city just dusting off its frontier image and intent on proving itself civilized.

During the trip to Tacoma a cool fall rain drove most of the other passengers inside. For a good while Martin stayed on deck, just under an awning, listening to the rain on the water as the boat stopped at various settlements. When it turned into Commencement Bay, Mount Rainier loomed above the landscape to the east. The city of Tacoma was dominated by buildings perched on hills above docks loading lumber for shipment around the world. One didn't quite notice this industry walking the shores of Seattle. The nearby forests had been denuded for decades, and logging operations had moved farther afield. But here trains pulled in giant trees from nearby foothills, and mills lined several miles of the bay's shore.

The College of Puget Sound was a long walk inland from Tacoma's municipal dock. James Slater had an office on the top floor of a brand-new building called Jones Hall. He was younger than Martin expected: tall, thin, and blue-eyed. He gave Martin a tour of his collection of local reptiles and amphibians, and described his plans to create a proper museum once a second building was built in the broad swath of clear-cut land to the south.

"Key to ensuring the future of a collection like this is to focus on local fauna, in my view," Slater explained. "Make it mean something to the city."

After the tour, they sat down in Slater's office and he handed Martin a copy of the description of the course on eugenics.

"Ordinarily I'd be teaching it," he explained, "but I've got to get out into the field next summer. Kincaid said you might be willing to branch out a bit now that you've got your sea legs teaching introductory zoology. You could combine your trips with work at the Experimental Station in Puyallup."

Martin read the course description as Slater talked:

Mental Hygiene and Eugenics—A study of the problems of mental physiology, laws of heredity, sex, and racial progress. The questions of responsibility for conduct; mental and nervous defects; crime and delinquency, racial betterment, the relative importance of heredity and environment in the development of the individual, are thoroughly considered. This course gives the sociological aspect of Biology.*

"Pretty broad sweep," Martin said, wondering how in the world he was supposed to define much less teach something called "racial betterment."

"Well, it's the kind of thing students want," Slater replied. "And we owe it to them to ensure practical meaning to their studies. Something that they can apply to their own lives. We had an essay contest a few years ago on social hygiene with the help of the Pierce County Social Hygiene Society. Most inspired writing I've seen from them in years. Much more so than when I ask them to write about lizards."*

Martin, who thought the main aim of scientific writing was that it wasn't, in fact, supposed to be inspired, said nothing as Slater passed him a list of exam questions he used in his own course:

1. What are some of the characteristics of the different classes of feeblemindedness?

2. What reasons can you give for the high percent of crimes in the US?

3. What is meant by the selective action of mortality?

4. What changes would you suggest be made in the State of Washington for eugenical reasons?

5. What change would you suggest be made in the College of Puget Sound for eugenical reasons?

6. If a trait is dominant, sex-linked, and lethal when homozygous or when not counteracted by a corresponding normal factor, what will be the sex ratio of offspring of mothers showing this trait?

7. What diseases are inherited?

8. Write on "Christianity (Christian religion) and Eugenics"*

Martin had no idea how to answer any of these questions, though he could guess how a biologist might tackle the third and the sixth. But when Slater asked, "What do you think, Sullivan?" he agreed to catch up on the latest eugenics literature before making his final decision. That, at least, could do no harm. Given the caliber of the biologists who supported eugenics, he thought, even hoped, he might be missing something.

He had another long walk heading north from the college campus to the address Trevor had given him for Reverend Gray's home. The two-story Victorian was nestled within a stand of huge trees that had somehow escaped the attention of the city planners and lumbermen. A woman opened the door. She was tall, almost Martin's height, which meant her brown eyes looked directly into his. When he introduced himself, and apologized for arriving so late, she spoke a brief "Oh yes, I remember," and motioned him inside, holding out her hand for his coat and hat, and turning away from him to place them on a set of hooks on the wall by the door.

Martin might have thought her very young, but for the style of her brown hair, which was long and twirled into a loose chignon at the base of her neck. One noticed long hair these days. Young women increasingly preferred short bobs that scandalized their Victorian elders. Martin decided that he liked the old-fashioned style. A line from one of his father's favorite reformers ran through his head: that it is the adorned, partially concealed body, and not the absolutely naked body, that inspires desire, and that was why humans had adopted clothing. To attract potential mates. Martin realized the trend of his thoughts just in time to avoid her noticing his gaze as she turned back to him.*

He followed her through a set of doors at the base of a staircase. It was a pleasant room lined so tightly with books that there was no space for paintings. A gramophone was tucked in a corner, in turn surrounded by books, and a Paul Whiteman Orchestra jazz tune sounded softly through the room. His guide brought him before an elderly gray-haired man, who stood with an unfocused gaze as they entered and she spoke. "You have a visitor, Father."

Martin stepped forward. "Hello Reverend Gray. The university sent me down about your collection. My name's Martin Sullivan."

"Professor Sullivan! Yes of course!" the minister replied, extending his hand.

He was taller than Martin, thin, with a great deal of thick white hair freshly trimmed, and very neatly dressed. He was even wearing an old-fashioned, dark brown cravat. It matched his eyes. Martin could now discern by the way he held out his hand too far to the right that he was completely blind.

"We had a note from Professor Kincaid that you would be coming," he was saying. "Sit down, sit down. Helen was just about to bring in the tea. Helen?"

"I'm here," said his companion. She had stopped the gramophone and was putting the record away.

"Could you bring another cup? And perhaps some of those delicious biscuits from this morning? Are you sitting, professor?"

As Helen disappeared, Martin said, "Yes, sir," and sat in a chair across from him apparently reserved for visitors. There was a third chair near the window, before a table full of books, papers, and a sewing basket. A dark blue velvet sofa. And a large specimen cabinet behind the sofa.

"Professor Kincaid said you worked at the museum at Harvard. I once knew that building like the back of my hand. Has it changed much, or can the ghosts of old naturalists still navigate it easily enough?"

"When were you there?" Martin asked.

"Let's see. Must be nearly half a century ago now. I was a student at Union Theological Seminary. A few of us used to visit the museum during holidays."

Martin knew enough of the landscape of the East's educational institutions to sort his host into the category of liberal Presbyterian, assuming he had had no subsequent conversion to either orthodoxy or heresy.

"I can't imagine it has changed a great deal," Martin replied. "The current director doesn't pay much mind to demands that the museum modernize. But it's still a place where good work can be done."

"You don't think it's behind the times?"

"Oh, I know it is," Martin replied with a smile. "I'm just not convinced that's a bad thing, if we can weather the current fashion that anything worth doing's done at a lab bench."

"It withstood the shift to Darwin's vision of nature." There was a mischievous bent to the minister's smile now. "Surely it can withstand the likes of a bunch of geneticists and Jacques Loeb."

"Maybe," Martin said. Something about Reverend Gray's smile and tone made him forget to feel nervous about getting into a debate about the most influential materialist in the country with a minister. "Sometimes the victory goes to the man willing to speak the loudest. I'm afraid our little entomological societies can't quite compete with Loeb's declarations that all of life can be explained with physics and chemistry."

"Yes, all sounds pretty revolutionary and important. But I found Loeb's work, a bit too, well, mechanistic, for my taste. He'd probably say I'm physiologically predetermined to dislike it. Darwin seems quite orthodox and old-fashioned in comparison, doesn't he?"

Martin laughed. "I've never heard Darwin described as orthodox."

Reverend Gray turned his head toward the sound of Helen's footsteps as she entered the room again.

"Helen, where is my copy of *The Origin*?"

She placed the tray of tea and biscuits between them on a small table.

"Just a moment," she answered. Having poured Martin a cup and handed it to him, she maneuvered between Reverend Gray's chair and the fireplace to the wall of books behind him. This was Martin's first indication that his host had a habit of appealing to the text concerned on any particular point of issue, and that Helen served as both his library assistant and reader. When she said, "Here it is," he asked her to read the last paragraph. Martin thought he knew this book well enough. But he had heard something unfamiliar by the time she was done.

It is interesting to contemplate a tangled bank, clothed with many plants of many kinds, with birds singing on the bushes, with various insects flitting about, and with worms crawling through the damp earth, and to reflect that these elaborately constructed forms, so different from each other, and dependent upon each other in so complex a manner, have all been produced by laws acting around us. These laws, taken in the largest sense, being Growth with reproduction; Inheritance which is almost implied by reproduction; Variability from the indirect and direct action of the conditions of life, and from use and disuse; a Ratio of Increase so high as to lead to a Struggle for Life, and as a consequence to Natural Selection, entailing Divergence of Character and the Extinction of less improved forms. Thus, from the war of nature, from famine and death, the most exalted object which we are capable of conceiving, namely, the production of the higher animals, directly follows. There is grandeur in this view of life, with its several powers, having been originally breathed by the Creator into a few forms or into one; and that, whilst this planet has gone circling on according to the fixed law of gravity, from so simple a beginning endless forms most beautiful and most wonderful have been, and are being evolved.*

"That's strange," Martin said when she had finished and handed the book to him. She sat down at the desk by the window as Martin glanced over the paragraph again. "I don't remember that bit about the Creator. Which edition is this?"*

"The sixth. I want to see how a man has replied to his critics, and whether he changes his mind, you see."

Martin turned the book over in his hands, and in that moment, the minister added, "My friend John Gulick loved those final lines."

"I assigned Gulick's work on geographical variation in Hawaiian snails to my students this term," Martin said. "The students were surprised when I referred to him as reverend."

"That's a minor tragedy, but seems to be the trend of things. John always insisted that the proof of Darwin's views lay in the power of his hypothesis to give meaning and harmony to a multitude of facts that otherwise remain isolated and unexplained. He used to lay out a long series of Hawaiian shells and show how individuals living on the trees of adjoining valleys differed in form and color from one end of a mountain range to the other. All his work in the field demonstrated the great variation in nature. He always believed that the argument for the being, power, and wisdom of God is only enhanced when one understands the birth of species, genera, orders, and classes as progressing according to a method established from the dawn of animate existence. Darwin's addition of that one word,

Creator, meant a great deal to him." He paused. "It's strange. How much we crave the imprimatur of the great masters, for things that aren't supposed to rely on the authority of any human being." And then he smiled. "Well, I have a habit of sermonizing. No doubt it's an atavism from my days in the ministry. Tell me. What does Professor Sullivan think?"*

Martin was caught off guard by the question. He happened to glance at the reverend's companion, who had been sitting at the desk by the window working on some task of her own. His usual pause turned to silence when he witnessed the concern in her eyes.

"Do you hesitate to speak your mind to an old man of the cloth who faces the imminent prospect of finding out who is right?" Reverend Gray asked.

Martin was spared from replying by Helen, who had noticed his glance.

"I'm afraid his silence is my doing. I remember the last time you debated these matters with Reverend Allen."

"Reverend Allen talks too much and I couldn't get a word in edgewise. Does our visitor look like he tends that way? He doesn't sound like it."

Helen smiled slightly as Martin looked at her again. "No. Not at all."

"Besides," the minister continued, "if he upsets me that just proves I'm still with you, with more life than were I to sit here benumbed into tranquility and peace by every visitor thinking the same as I do. I know you understand, Helen. Can he proceed?"

Martin struggled for a moment to remember the question once Helen had given permission by returning to her work.

"I usually don't know what to say on such matters," he offered finally. "My father raised me as a firmly committed agnostic. It tends to justify ignoring things on which other men speak with certainty."

The minister smiled. "Odd phrase that. Firmly committed agnostic."

Martin nodded, and then remembered that his host could not see him do so.

"Yes. I'm afraid it isn't very helpful at times. Not when my students are afraid to tell their parents what they're learning in their biology courses."

"That's a vestige of the ministerial foundation of your profession pulling at your secular robes. It does you credit that you feel the difficulty."

"Maybe," Martin replied. "Feeling the difficulty doesn't make me any better at solving it."

Then, because of the silent, patient way the minister waited for him to say more, he spoke of Pete's dilemma and how unprepared he had felt to aid a student trying to figure out what were the implications of science for faith. Which meant he had to speak of Connor's death and how he'd not known what to say amid the conflicting pulls on Pete's conscience. The only words that came into his head, he confessed now, were Ingersoll's warnings that Nature knew nothing of prayer.

It was an uncharacteristic confession. In the midst of speaking, he realized how much he'd shared without the trembling setting in. He glanced at the reverend's companion. Her head was bent over some task at her table and he could not see her face. But she gave no indication of her previous anxiety. She gave no indication she was listening at all.

Reverend Gray had listened closely in silence. Now he asked, "What did you say?"

"Something about there being Christians who believe in evolution. But then I read Bryan's book. He makes short work of that claim. Says it isn't sufficient to say that some believers in Darwinism retain their belief in Christianity. Some survive smallpox. And, as men and women avoid smallpox because many die of the infection, so they must avoid Darwinism because it leads many astray."

"Only if they make it so!" Reverend Gray said, his brow furrowing. "Which is a pretty piece of irony, given Bryan's insistence that Darwinism is to blame for men and women's abandonment of Christianity!"

"You don't agree with him, then," Martin said.

"No, I don't. Young men and women are fascinated by science. They believe in its power to alleviate human suffering. In germ theory. In the power of a steam engine. In the possibility of human reason. If they are taught that if God did not make us by fiat then we are nothing but beasts, Bryan's campaign will reap a terrible harvest. Students who are told that they must give up modern science in order to retain Christianity will, I wager, abandon Christianity."*

"I wish Pete's father could hear you say that. He might trust your assessment a bit more."

"Oh no, professor. I'm even more dangerous than you. Bryan's been a bit distracted by this campaign against biologists, but we're the real trouble. Unorthodox ministers. Those who call themselves liberal or modernist Christians. I suspect Pete's father knows that." He paused. "Still. Choosing a career in biology isn't going to keep the boy out of our clutches. Or yours." He took another deep breath, and smiled as he said softly, "A bit ironic, isn't it? You spend your days ordering life. Yet you are blamed for striking all order, progress, and purpose from the universe."

Martin nodded. "And yet some look to us to establish a new order and new salvation. I fear both positions are too heavy a burden for us to bear wisely. We don't seem to be much good at either."

"Ministers aren't always much better at the tasks. Of course, I always blamed that unfortunate fact on our humanity, rather than our beliefs as Christians. Just as I suspect naturalists' missteps are due to the fact they're human beings, rather than any particular ideas they may hold about their little specimens and the origin of species."

The word specimens reminded Martin of the collection he had come to evaluate. The fact he had momentarily forgotten it astonished him. He apologized and Reverend Gray laughed, for he had forgotten the collection as well.

"Helen, can you bring him a few trays? Set them on the sofa, so the professor can take several in at once and get a feel for the collection."

Helen left her desk by the window, went to the large cabinet behind the sofa, and began pulling out trays. Martin knelt down and looked at the specimens that she laid out before him on the sofa, speechless. One could tell when a naturalist's love for insects had been disciplined into a scientific approach to nature, and not just because the creatures had been asphyxiated with laurel vapor. Many amateur naturalists ordered their collections by color and size, creating aesthetically pleasing trays of very little use to entomologists trying to trace evolutionary relationships. But these trays were filled with thousands of specimens of lady beetles from throughout the West Coast, each specimen neatly labeled with a detailed locality and date, and arranged in a manner that let one see the geographical variation at a glance.

"How many trays are there like this?" he said finally, as Helen brought two more from the cabinet. She was very careful with them, gently wiping each glass with a damp cloth before placing it alongside the others.

"About ninety, I think. There's another cabinet upstairs."

"Good grief," Martin murmured under his breath. He looked up at her. "Is there a master list?"

She smiled slightly and gave a small nod in the reverend's direction.

"In his head. I've been trying to make a paper copy from the specimen labels. He knew you would want one."

"I had no idea the collection was so large," Martin said. "An inventory list will take some time. But we'll need one for the appraisal. I can help if you like."

"We can give you as much time as you need," Reverend Gray spoke up.

Helen brought a few more trays, and Martin soon lost all track of time as he asked Reverend Gray about various specimens before him. He'd have missed the last boat back to Seattle if Helen hadn't asked him at some point if he was staying in Tacoma for the night. She had obviously glanced at the clock. Martin apologized and stood up when he realized what time it was.

"Can't we call up Charles to give him a ride?" Reverend Gray said, and then turned in the direction of Martin's voice. "We've got a telephone now. In the hallway. Did you see it?"

Martin smiled, charmed by the boyish enthusiasm in his host's voice. "No. I'll look on my way out."

"We best not impose on Charles too much," Helen said. "Everyone is asking him for rides these days." She looked at Martin. "You're welcome to come again."

Reverend Gray seemed delighted by the idea of another visit and stood up as though that settled the matter. As he held out his hand he added, "How well do you know this book that Bryan insists will result in the downfall of civilization?"

"*The Descent of Man*? Not well at all," Martin answered. "Though I've read *On the Origin of Species* half a dozen times." He smiled. "The first edition, that is. But *The Descent of Man* always seemed so far from my own work."

"Until now?"

"Yes. Until now." Martin smiled again, and wondered whether his host could hear the smile in his voice.

"Well. Since you're coming again, what do you say we read a few chapters together? It seems we might both benefit from thinking a bit more on the matter." He smiled. "We could be good checks on each other's wish to read the book as we like."

Martin hesitated, for clearly his host's idea must involve the time of a third person. He glanced at Helen, who had retrieved his coat and hat and was standing at the entrance to the hallway. She was watching them both, and she smiled slightly as she acknowledged his gaze, but she gave no indication she approved or disapproved of the plan.

"All right," Martin replied finally. "I'll get myself a copy tomorrow. Last edition?"

The minister laughed. "Of course!"

Martin said goodbye and followed Helen to the door.

"I'm sorry for staying so late," he said as he put on his coat and hat. "I lost track of time."

"It's all right," she answered. "I'm sorry for halting you from speaking. I didn't mean to."

Martin smiled at the idea he might need to be halted and said good night.

As he traveled home, he tried to make sense of the rapidity with which he had agreed to a proposal that would involve sharing his thoughts on topics he tended to avoid, with a man, a minister no less, whom he had just met. He wondered whether a Freudian would diagnose the decision as a subconscious means of soothing his conscience in the wake of a life-time of his dad's "You need to look up from your bugs now and then, Marty!" But in the end he decided he had agreed because he liked Rever-end Gray very much.

—

Trevor looked at Martin in astonishment when, upon entering the collection room on Monday morning to ask how the trip to Tacoma had gone, Martin confessed he hadn't looked at more than a few trays and would be going back again the following week.

"Goodness, Sullivan. Was there a pretty daughter in the house or had you been drinking?"

Martin laughed. "Worse. We got into a discussion about God and evolution."

"Oh dear. That really isn't like you. Well. What was the outcome?"

"We decided we both need to do more research."

"Ah. There's Professor Sullivan. Found a kindred spirit then?"

Martin smiled but said nothing.

"And Slater?" Trevor asked. "How'd that meeting go?"

"I promised to look into it. He said he didn't need a decision until spring."

Trevor chuckled. "Sounds like you've got some homework to do. Can't wait to hear what your dad thinks."

Martin hadn't planned on telling his dad about either set of home-work and felt a bit guilty about that fact. But it turned out Will was too distracted by his own affairs to interrogate his son about what he was doing with his days. Her name was Rebecca Shelton, and Will had met her on the Tulalip Reservation. He brought her to Martin's office in the collection room one night when Martin was working late. Everyone else in the science building had gone home, so Martin was surprised to hear voices, including a woman's pleasant laugh. Then he recognized his father's cheerful tone and looked up as the footsteps neared. Martin came from behind his desk as they entered the room. When Will introduced Rebecca the edges of her brown eyes wrinkled as she smiled. She was older than the women Will had spent time with since Mary's death. Her gray-tinged, black hair was neatly bobbed. She listened as Will tried to explain what Martin did with his days by giving her a tour of the specimen trays lying about.

"How am I doing?" Will said, glancing up at Martin, who sat on his desk, watching them.

"Just fine," Martin replied with a smile.

Rebecca looked at the dead insects very closely as Will talked. When he was done, she looked at Martin. "It's very still in here," she said.

For a moment he couldn't figure out whether she was referring to the insects immobilized within the trays, the air in his office, or his own heart.

"I told you he needs to get out and about more," Will said.

"But he feels he must be here," Rebecca replied. "And what's this? What are you doing here?" She had moved to his microscope and was looking down at the papers where he had set down his pen.

Martin helped her focus the microscope so she could see the specimen beneath the lens.

"This is a group of weevils that's given taxonomists a lot of trouble. It's been described as two dozen different species. I think it's actually just three or four, with a lot of geographical variation. So I'm trying to sort them out. Clarify the names."

"It is a lot of death for a few distinctions," she said quietly as she turned from the microscope and looked again at the trays.

"Yes," Martin replied. "This one here's described as a separate species. I'm pretty sure the series shows it doesn't deserve the title. See here? They grade into each other. With enough specimens they're all collapsed into one. The gradation is a good clue that they can interbreed."

"Then you're getting rid of the names. You aren't making species."

Martin smiled. His dad must have told her about Ben.

"In a way the names are just holding places. Important thing is to be willing to let the next specimens upset the boundaries the name was meant to capture, and start over again. If the specimens tell you to."

Rebecca looked up at him. "You must be a good listener."

"I try to be," Martin answered. "Isn't always time, though."

"Still. Some don't seem to even try." She moved away from the trays and sat down beside Will as she continued. "The historians from this place come to the reservation and want to know the names of everything. They ask about the big mountains, islands, rivers, and lakes. What do you call this? What do you call that? But my mother's people don't have names in that way. We have names for the places where the camas grows, where one can slide cedar planks easily down a hillside, where trails cross, and good landing sites for canoes."

"Who was that anthropologist who came last month?" Will said with a chuckle. "The one who got all frustrated and insisted the lack of names for big features of the landscape reflects Indians' primitive ways of thinking? I'd say that's pretty primitive thinking on his part."*

"At least Dr. Gunther's different," Rebecca replied. "She just wants to know the stories that come with the names. She understands it's the stories that matter."

"I heard there's some fight going on about whether Mount Rainier should be renamed Mount Tacoma on account of that being its Indian name." Will was thumbing through one of Martin's reprints as he spoke, but didn't find anything of interest so he tossed it back on the pile. "You heard about it, Marty?"

But Martin waited for Rebecca, because she smiled broadly at Will's question.

"Oh, that's all quite silly. They're saying Tacoma is the Indian word for a woman's bosom."

"Sounds like something a white man with a complex might imagine." Will laughed.

"Some rich city ladies tried to get the Puyallup to join the fight," Rebecca explained. "One of their leaders, Henry Sicade, asked why his people should join in a dispute over the name of the mountain. Asked those women to put themselves in an Indian's place, undergo all we have gone through, then imagine how readily a people will fight to retain an aboriginal name after losing all their lands." She paused before adding, "We don't argue over the names of rivers, lakes, or mountains."*

"No wonder there aren't any Indian taxonomists," Will said. "What do you think, Marty? Bit frightening after spending your life as though the names matter, isn't it?"

"You said the stories are what matter," Martin said, ignoring his father's teasing.

Rebecca nodded.

"Tell Rebecca one of your tall tales," Will suggested. "About these little beasts."

Martin would have ignored his dad again except for the interest with which Rebecca turned to him and said, "I would like that." So he went to one of the cabinets filled with trays he used for teaching. He pulled out a tray of specimens and placed it on his desk before her.

"This one is called *Dendroctonus brevicomis*. An entomologist named LeConte gave it that name in 1876. This one is *Dendroctonus punctatus*." He moved his finger. "*Dendroctonus similis*. Placing them all in the same genus—*Dendroctonus*—indicates they all came from a common ancestor. That they're all related."

"That isn't obvious?" Rebecca said softly.

"Well, yes," Martin replied with a smile. "I suppose it is. But Linnaeus, who came up with this naming system, used the word relationship to mean patterns in the mind of God. He didn't think they were related because they shared a common ancestor."

She looked up at Martin and smiled. "So you are like Adam. In the Garden of Eden."

"Undermining Adam's more likely," Will interjected. "Adam's species didn't change one into another. God fashioned them just so."

Rebecca looked at him. "You don't know that," she said, and then turned back to Martin. "Your father always has an old man with a white beard in his head." And then she added, looking at the specimens, "But one might imagine a great Spirit, Creator of all things, whose handiwork is in the forests and the streams, in the mountains and the seasons."*

Martin glanced at his father. This wasn't normally a kind of talk he tolerated well. But Will was looking at Rebecca with admiration rather than disdain.

"Now. Why these?" Rebecca said, gesturing to another tray on Martin's desk.

"I don't usually study this group. Timber men asked me to look into them. Apparently there's an outbreak on Vancouver Island and they're afraid it'll spread."

Martin could feel his dad's indignant stare as he spoke. "And you said yes?"

"Didn't have a good excuse to say no, Dad. Best they ask an entomologist to get the identifications right before they start fumigating entire stands with Paris green for no reason."

"Well they'll cut 'em all down soon enough. I don't think they need your help turning a profit."

Martin said nothing and began packing up his satchel.

Rebecca joined them at their apartment often over the next few weeks. Will would cook up a stew for a working dinner, and then they sat for hours, heads bent toward each other, looking over a pile of papers. He had offered to look over Rebecca's will, to help her ensure things passed rightly to her own children, and that led to other stacks of papers belonging to close relatives involved in land disputes.

Martin suspected an attraction pretty quickly. His father loved women willing to speak their mind, and Rebecca, once she was sitting down before a stack of documents, and Will began his questions, did so with conviction. Martin learned that her dead father was one of the early settlers, and her mother the daughter of a leader of the Duwamish who had helped those settlers survive.*

"My grandparents married outside the tribe for good reason," she explained. "You could walk through another people's territory if you were bound by marriage. Maybe at one time the whites believed that too. Until they didn't need to ask for permission. And could make their own rules. They made laws against marriages like my parents'. First against Indian ceremonies. Then against whites marrying Indians at all."*

"Those still aren't on the books here, are they?" Martin asked. He knew there were laws against miscegenation in other states, and he knew some biologists defended those laws because that fact had inspired so many of his dad's "What are your damned colleagues about, Marty?" But he'd not heard of such laws in Washington State. And surely his dad would have mentioned them.*

She smiled. "Oh no. Reason or conscience prevailed in the end."

"Or all those men realized how many of their heroic pioneers' children would be rendered illegitimate," Will added.

Will picked up a few of the papers Rebecca had laid out before him on the table.

"It's a lucky thing. This looks all right, Rebecca. But I think it best you transfer the boys to a public school just the same. Best they know how things work outside the reservation if they ever have to stand before a judge with these papers."

After she had carefully placed the papers in an envelope, Will walked Rebecca to the house of a cousin who lived downtown. When he returned and sat down opposite Martin at the table they used for everything from meals to bills to darning socks, Martin could tell he wanted to talk, so he put his own reading aside.

"The children, Marty," he said. "The children. There's a boarding school on the reservation. The poor kids aren't allowed to speak their own lan-guage. Get caned if they do. Rebecca had to convince the school to take her children, since they aren't full Indian. Thought it would be best for them."

He shook his head, as though that was all that was needed to indicate she was wrong.

"Her eldest boy was at Cushman Indian School in Tacoma. Bureau was bringing kids down from Alaska to keep the school's quota up. Can you imagine, Martin? Alaska! Taken from their families so the white man can eradicate their language and beliefs. That's the word the bureau uses. Eradicate." Will gestured at the reprints of entomology journals beside Martin's plate. "As though they're insects!"*

Will paused for a moment, lost in thought. Martin said nothing, but he did not return to his reading.

"They lost him during the flu epidemic. Poor husband had a breakdown after that. Doctors say he succumbed to drink because Indians have weak willpower. But, for God's sake, you show me a man who can claim with certainty he wouldn't succumb under the same circumstances?" He sat back in his chair and then worked out his thoughts aloud. "Her boys will have a better chance at the public high school. Won't be under the thumb of an Indian agent who insists they listen to fire-and-brimstone sermons every Sunday."

Martin looked up at his father. That was not the reason he had given Rebecca.

"What does Rebecca think?"

"Well, I think she's all right with the idea."

Martin saw the wheels of his father's "on the other hand . . ." churning. The moment of doubt would not prevent his dad from acting in the end, but he nodded, chastened by Martin's question. "I'll ask her again. Just to be sure."

Chapter 8

Within moments of Martin's second visit, Reverend Gray insisted that Martin call him Josiah, so Martin asked Josiah to forgo the "Professor Sullivan."

They settled into a pattern of work that would guide Martin's Friday evenings for weeks: a few hours on the collection list and then an hour discussing the week's assigned reading from *The Descent of Man*. On that second visit, Josiah asked Martin if he could debate while working on the specimens, and Martin laughed.

"Not usually."

"We best work on the collection first then, so we don't forget," Josiah offered.

Martin's admiration for Josiah's work grew with every tray of specimens. Sam Henshaw had often lamented the inability of taxonomists to study evolution because they continued to collect as though species were created by God: one male, one female, two perfect specimens of each, job done. To truly serve as sites for the study of evolution, he urged, museum drawers needed to reflect the variation in nature: one hundred specimens of each species might not be enough! Looking over Josiah's collection, Martin estimated there were three hundred specimens of one species—*Hippodamia convergens*—alone.

"Did you ever do any microscopic work on these?" he asked during his second visit. He'd have been more specific, and asked if he'd looked at the genital armature, but Helen was kneeling nearby, and Martin had no idea whether she knew that studying the insects' vaginal plates and penile armature was a basic requirement of determining the species correctly.

"Oh no. I hope someone will. There are a few species I'm pretty certain aren't species at all. But I never had the time or equipment to go into the question aside from the hunch." He smiled. "And then, well, I've got plenty of time now, but I'm not much good with a microscope."

"I can bring one from the university and check your suspicions if you like," Martin offered.

When they did set aside the specimens and turned to *The Descent of Man*, Martin was surprised to find Josiah full of nothing but praise for the first chapters. He spoke with admiration of the rhythm by which Darwin asked, after placing a whole series of facts about human anatomy and physiology before the reader, which theory explained them best, "descent with modification" or "the theory of special creation."

He requested that Martin reread certain passages and then asked what Martin thought of Darwin's reasoning. So Martin read a long excerpt of Darwin's evidence of the similarities between monkeys, apes, and humans aloud.

These monkeys suffered also from apoplexy, inflammation of the bowels, and cataract in the eye. The younger ones when shedding their milk-teeth often died from fever. Medicine produced the same effect on them as on us. Many kinds of monkeys have a strong taste for tea, coffee, and spirituous liquors: they also, as I have myself seen, smoke tobacco with pleasure. Brehm asserts that the natives of north-eastern Africa catch the wild baboons by exposing vessels with strong beer, by which they are made drunk. He has seen some of these animals, which he kept in confinement, in this state; and he gives a laughable account of their behavior and strange grimaces. On the following morning they were very cross and dismal; they held their aching heads with both hands, and wore a most pitiable expression; when beer or wine was offered them, they turned away with disgust, but relished the juice of lemons. An American monkey, an Ateles, after getting drunk on brandy, would never touch it again, and thus was wiser than many men. These trifling facts prove how similar the nerves of taste must be in monkeys and man, and how similarly their whole nervous system is affected.*

Josiah was laughing by the time Martin finished the passage. "It's quite a different argument than the parade of skulls brought out by your fellows and placed all in a line to prove humans are apes, isn't it?"

"Yes, it is," Martin replied. "It's a better argument."

There was very little to divide them at first. Josiah stuck resolutely to his stance that Darwin's work raised no new problems for the thoughtful Christian.

"Bryan argues that the belief that man has developed from primordial forms via fixed laws of evolution somehow vitiates faith in a Creator," he said. "But that makes as much sense as saying the development of each individual from an embryo via the fixed laws of development disproves God. Even if we were made by a gradual process, so long as God is the creative power, surely our relationship to our Creator remains unchanged."*

That was something he might have said to Pete's dad, Martin thought. It was helpful testimony to have if the subject ever came up again and he had to take on Reverend Harrison's assumption that evolution and Christianity were in conflict.

On his third visit he brought an old microscope he'd found in a wooden box in the Zoology Department storeroom.

"You best set it up here," Helen said, clearing a sewing basket off her desk. "This corner has the best light."

"But that's your workspace," Martin protested. "I'll be fine over here."

He took the microscope to the large bay window that looked out over the front garden toward the street, where he'd have to stand as he worked. Within an hour he had a stack of trays beside him that contained lady beetles that Josiah, from memory, pronounced as only tentatively identified.

Those Friday afternoons spent over Josiah's beautiful trays of specimens, amassed for no other reason than a wish to trace the paths of insect diversity, became a much-beloved refuge from the demands of teaching that fall. Though it was his second year in the classroom, Martin still constantly second-guessed himself after professing about the natural world before his trusting students. They had to be examined on something during finals week. But here, with Josiah sitting near, quietly cheering him on as he tried to figure out the genera to which dozens of undetermined species belonged, he could go slow, withhold judgment until he was certain, and be guided solely by curiosity about what was more closely related to what. He could once again sit before a collection of insects with no bearing on man's understanding of either economics or God, and feel he was doing good. He soon grew to love these evenings at Josiah's fireside, even the discussions of *The Descent of Man*, and to feel at home there.

Except for one thing. He found it difficult to determine how, or even whether, he should speak to Josiah's daughter. Well trained by his father, Martin struggled to get used to the fact that Helen did not join their discussions of *The Descent*, though she had read every line aloud to her father. She rarely spoke, even as they worked through the trays while sitting on the floor together. Martin checked each specimen label and read it to Josiah, and she transcribed the information on the labels into a notebook.

She had to sit quite close, so she could read the label as well and keep track of which specimen he was looking at. Within their first moments of working together Martin observed her hands as she quickly and neatly wrote on the specimen list, *#302. Coccinella trifasciata. Mitchell Bay, San Juan Island, April 2, 1901.* Over years and years of observing the appendages of weevil specimens he had learned to notice the slightest variation in color and form of their tiny claws: all the same basic pattern, yet with minute differences if one looked closely. So he noticed the shape of her fingers, the turn of her wrist as she pressed one hand against the paper as she wrote, and the graceful way she moved the pencil.

He thought she might speak more when, on that third visit, they all forgot the time and Helen suggested he stay for dinner, but her words were still limited to matters concerning Josiah's comfort. The meal was fresh salmon, and Josiah said that it would be that way for just a week or two more, when the runs stopped. It was delicious, and Martin said so. He sensed from a slight turn of her lips that she did not like the attention. He wondered, then, whether his dad would have been able to charm her into speaking.

Having caught an occasional glance of anxiety in her eyes when Josiah became tired, Martin took to relying on her for notice of when their conversations must end, for Josiah seemed loath to admit the defeat of fatigue. They were almost imperceptible signals: the silent setting aside of her pen and paper, or the quiet shelving of a book she held in hand, but he soon developed a good ability to sense the gestures.

She never gave another indication of fear that certain topics might upset her father, even when, as the weeks passed and they read further into *The Descent*, Martin noticed a distressed bent to Josiah's brow. They had come to the section in which Darwin added "Belief in God" to the long list of supposedly unique, human traits that were in fact shared with other animals. Darwin proposed that man's belief in God arose from instincts that some animals possessed as well, albeit in a lower form: imagination, memory, and association. Over time those traits had led, in human beings, to the impulse to explain experience in terms of cause and effect. But once again, Darwin argued that the difference was one of degree, not kind.

"I do fear that so evolutionary an account of belief in God cannot help but undermine that belief," Josiah said. "For some."

Martin replied by reading aloud the caveat with which Darwin had begun the section:

The question is of course wholly distinct from that higher one, whether there exists a Creator and Ruler of the universe; and this has been answered in the affirmative by some of the highest intellects that have ever existed.

Josiah frowned.

"That shows he believes nothing. It is a passive sentence. He does not take a stand. He finds the roots of the religious faculty in the same sentiment that creates a dog's love for its master. Then he gives a footnote!"

"He is pointing out that the claim that a dog looks on his master as on a god is Braubach's, and not his."

"It is his, Martin. It is his. He is hiding behind the footnote."

"But what alternative does he have?" Martin countered. He had already astonished himself by pressing Josiah on certain points. "What would the book be if he offered a stronger statement that God exists? His point is that all the wonderful powers that seem to make man so unique could have arisen via evolution, rather than direct endowment. No exceptions. You conceded the endowment could still be distant and in the cards from the beginning. Doesn't that mean an explanation that derives religious beliefs from our animal past should be no more threatening than the derivation of our rudimentary tails?"

Josiah smiled. "But you don't believe that, Martin."

"No, I don't," Martin replied. "But my disbelief does not arise from a scientific explanation of belief in God."

Martin had said he could not usually debate and work on specimens at the same time, but faced with Josiah's gentle yet pointed queries, he had discovered he could do things in this house that he had always avoided. There was no trembling. No physiological effect. Eventually, he even got used to following Josiah's lead in occasionally forgetting Helen's presence, and that she, too, had read every line they discussed.

Sometimes, when he remembered, he wondered whether Josiah could tell when she was in the room. She moved so quietly. The blind were supposed to have heightened powers of perception, but he could tell from the dates on the specimens that Josiah's blindness had come late in life. He had his answer when Josiah suddenly paused their discussion of Darwin's account of the origins of belief in God and leaned toward him.

"She trusts you."

Martin looked up from his book at the "she" and noticed Helen had left the room.

"I'm glad to hear it," he replied.

"We don't agree on many things. But there's little so dear as a fellow human being's worth. The other things don't in fact matter so much."

Martin had sensed no conflict between them. The disconnect between Josiah's hint of disagreement and Helen's constant attention and care was puzzling. But he was glad of this testimony that she did not mind his presence. Especially when they proceeded to the chapters in which Darwin addressed the origin of morality and conscience "exclusively from the side of Natural History," and the gulf between how he and Josiah read the book widened.

"Do I have this right?" Josiah questioned as they came to Darwin's attempt to derive humanity's moral sense from its animal ancestors as a means of collapsing the boundary between them. "That in animals who derived some benefit from living in groups, natural selection produced the ability to feel anxiety when separated from one's fellows, and that that anxiety might even be called love? That the basis of all our greatest tales of loyalty, courage, and love affairs, is this ancient social instinct?" Martin assented to the summary, and Josiah exclaimed, "He has gone too far, Martin . . . He has gone too far!"

"He went so far in *On the Origin of Species*," Martin replied, "when he wrote that his theory would throw light on psychology and the origin of man. He knew it must encompass everything. Every behavior and every characteristic. Including conscience. Including love. But I don't think he believed that their evolutionary origin strips either conscience or love of their power."

"Then he didn't think hard enough," Josiah pressed. "If our moral sense evolved purely from some animal past, who is to say whether we must obey it? Who is to determine which rules are valid? And what kind of anchor is this for determining what is right?"*

"Is the presence of that dilemma a fair argument against the validity of the theory?" Martin replied.

Josiah pressed his hand to his lips. After some moments, he spoke slowly, as though he was working out his fears as he did so.

"But if our conscience is simply the impulse of our social instinct in conflict with our selfish instincts, then regret is merely a feeling of unease from giving in to one instinct versus the other. Our moral sense commands no special respect, arises from no unique authority, and admits no external standard of right or wrong. Indeed, Darwin believes that many of the rules humanity has laid down as moral are merely the result of prejudice and superstition."

"As do you and I, surely. Human sacrifice. Slavery. Celibacy," Martin answered. "But he is grounding our capacity to develop moral rules, rather than the rules themselves, in our evolutionary past. Perhaps in doing so, he can account for why men are led astray so easily."

"But do you trust human beings to decide rightly, when Darwin gives them leave to decide which rules are the missteps of instinct and which are right?" Josiah pressed. "Do you think men can accurately determine which ones might be dispensed with? With no external sanction as a guide for what is good? Surely each man may be too weak to prevent a whole host of illusions from declaring his verdict has been rational and logic is on his side, when in fact he is justifying his original desires. Men may decide that might is right, because they wish it to be so."

Martin nodded, though Josiah could not see him.

"I agree with you. Wholeheartedly. But consider what he has just done. He has rooted our sympathetic feelings, including love, not in some divinely conferred gift that separates us from all of nature, but *in* our nature. Sympathy is what allows us to cohere in social groups. It's the first impulse to benevolent actions and compels us to imagine what the pain of another feels like. Compelling us to relieve that pain. He says man's evolution culminates in the Golden Rule, 'As ye would that men should do to you, do ye to them likewise.' This means the rule is rooted within us, instinctually, from birth. That is surely not a message that might is right."

Josiah smiled. "You are now arguing for the theory on the ground it will lead to good."

"No. I'm not," Martin replied firmly. "I'm not sure he's right. But I am arguing that it need not lead to evil. And that to say evolution leads to a might-is-right ethics leaves out a fundamental part of human evolution. Cooperation and sympathy are just as much a part of our animal heritage as competition and struggle. And if that's true, which one a man chooses as a guide or justification of his own behavior probably has very little to do with what nature tells him."

Josiah sat back in his chair. He had leaned forward as though straining to understand Martin's view. Sometimes, Martin had the sense that he was trying to see, so that he could read in Martin's eyes what he could not discern in his voice. At other times Josiah seemed to see much more than Martin.

"You are very consistent, Martin, I'll give you that," he said after a moment. "In your sympathy with Darwin's effort to bring each aspect of human behavior within the purview of a purely natural explanation, no matter what the cost. Your stance fits your profession, and its rules. But I must pay attention to the meaning, or lack of meaning, such efforts pose for those trying to find a moral compass in the world."

As they read further, Josiah alternately rebelled against certain passages and nodded his head in vehement agreement with others. Martin eventually discerned a consistent trend to the passages that inspired his friend's admiration. He loved Darwin's statement that "if man had not been his own classifier, he would never have thought of founding a separate order for his own reception." He praised the strenuous demand that man look at nature, and himself, with the utmost humility. But he despised the passages from which certain men might derive pride, especially Darwin's use of the words "lower" and "higher" for human races or "savage" and "civilized" to indicate a group's status on some scale of moral and social development.

When they came to the chapter "On the Development of the Intellectual and Moral Faculties during Primeval and Civilised Times," he asked Martin to read two brief passages aloud:

At the present-day civilized nations are everywhere supplanting barbarous nations, excepting where the climate opposes a deadly barrier; and they succeed mainly, though not exclusively, through their arts, which are the products of the intellect. It is therefore highly probable that the intellectual faculties have been mainly and gradually perfected through natural selection.

At all times throughout the world tribes have supplanted other tribes; and as morality is one important element in their success, the standard of morality and the number of well-endowed men will thus everywhere tend to rise and increase.

Josiah shook his head as Martin finished reading.

"And thus the extermination of those he calls savages is an inevitable decree of Nature, rather than the terrible, avoidable, outcome of the sinfulness of civilized man's cruelty."

"Men might read the text thus," Martin said slowly. "That does not make them right. Think of what he was trying to do, Josiah. He rebelled from his fellows' desire to make human beings exempt from natural law. Because doing so made no sense of the world. Because the suffering, disease, and misery they experience, yes, because men are tribal and hate each other, seems to indicate quite clearly that they are not an exception, and men best wise up to the fact, if they are going to understand themselves. He is not trying to justify such things. He is trying to provide a better explanation than one that holds a benevolent God created a world in which such evil exists."

Josiah fell silent for a moment, and Martin wondered if his friend had any suspicion how unlike it was for him to speak at such length, much less about another man's motivations. Even one who was dead.

"So he has not, in fact, removed his heart from his attempt to understand the world?" Josiah said finally.

"No. How could he? He has left his beetles and barnacles aside and is trying to understand his own kind."

"Yet nowhere does he demand that men protect their fellow human beings from being extinguished from the earth by the march of progress."

Martin shook his head, in an autonomic rebellion against criticism of Darwin on these grounds, though Josiah could not see him.

"He was a naturalist, Josiah. Not a politician or a moralist. He recorded what he observed. He tried to explain it. He is trying to convince his readers that his theory accounts for the observations, and that the phenomena observed is evidence of his claims. That does not mean he approved of anything his fellows were doing."*

"A man might disapprove of a war yet be just as complicit as a man firing shells on the battlefield," Josiah replied firmly.

"But what do you want him to have done?" Martin countered, "Signed on to a bunch of government commissions? I suspect most men cannot serve two mistresses safely. Not even Darwin. Had he been distracted by statements about the implications for his fellows, he would not have been doing science."

"Are we talking about Darwin? Or Martin Sullivan?"

Martin fell silent. And then he shrugged.

"Both, I imagine," he replied, smiling to himself. "This is precisely why I stick to beetles. No matter what I claim, if I am wrong, at least it's of little import to human beings."

Josiah let out a quiet chuckle.

"You might tell yourself that, Martin. For a while yet."

—

A few days later Will returned home to find Martin reading *The Descent of Man* on the deck. Martin was unsuccessfully warding off the evening's late fall chill with a blanket but refused to give up the view of Mount Rainier bathed by a setting sun. His father found him staring south, the book forgotten on his lap, and his fingers pink from the cold.

When Martin finally divulged what he was doing with this Friday evenings, Will had been both astonished and delighted at the news Martin had agreed to read *The Descent of Man* with a minister. "You best be on your guard, my son," he said with a laugh. "Strange things can happen to quite rational men in such company." It all proved a good excuse for sharing more newspaper clippings about Bryan's campaign.

"Ammunition against your reverend," Will said now, tossing some pamphlets in Martin's lap.

"You and Reverend Gray actually have a great deal in common, Dad,"

Martin countered, setting the pamphlets aside. "You should go down with me some time. I suspect you'd like him a great deal."

"I'm afraid I've lost all my patience with ministers of any stripe, Martin. I've seen the damage they've done on the reservations. The missionaries taught the Indians that if they accepted the white man's God, they would have everything good, from picket fences to refrigerators. What did their young men gain from Christian civilization? The chance for an all-expenses paid trip to the battlefields of Europe to be blown to bits by progress."

He took a deep breath, as though to rid himself of the image created by his own words.

"But isn't it the missionaries, Dad, who took a more optimistic view of Indian capacities?" Martin said. "For morality and civilization? However misguided their views of what constitutes either."*

"You think that makes up for those who insisted the struggle between races was designed by God to ensure the progress of civilization? No matter how many must be slaughtered? There's an historian in Oregon who says God oversaw the destruction of the Indians by disease so that the pioneers would be spared a stronger foe. How's that for Christ-like compassion?"*

"That doesn't make Christianity wrong, Dad."

Will sat very still, deep in thought. Then he suddenly smiled, as though at some memory.

"By God, you're your mother's child."

Martin didn't know what to say to that, so he changed the subject.

"What are you in town for this time?"

"Two of Rebecca's cousins have been arrested for fishing in the bounds of the reservation. Within the boundaries!" He banged his fist on the railing. "Judge with brains and a conscience threw the case out, but the fish commissioner's threatening to arrest them again. Get this. Commissioner's name's Darwin. Irony there somewhere. He's threatening to enforce state fishing rules on the tribes. The treaties guaranteed them fishing rights in perpetuity. I'm gonna see if I can't get one of the bigwig city lawyers to take the case."*

Will was obviously upset so Martin laid *The Descent* on the table beside him.

"People complain the Indians are wardens of the State, but the white man takes all his natural means of support. And what does he tell the Indian who has the courage to cry foul?" He gestured at Martin's copy of *The Descent*. "A bunch of scientific-sounding garbage about natural laws and the survival of the fittest. It's the heartless vision behind all our Indian policy. Not active extermination, mind. We're far too civilized for that. No. Just calm annihilation of everything Indian in the name of bringing them into civilization. Because otherwise their extinction can't be helped. And nature's laws must not be opposed."*

Martin shook his head. Darwin's words were fresh in his mind and he had something to say in response to his dad's outburst.

"Darwin didn't claim that the fact intellectual or moral advance arose from competition between tribes justifies anything, Dad. One can read a message of humanitarianism and justice to Indians, and anyone else, from these pages as well." He turned the pages toward a bookmarked passage as Will watched him, a bit astounded he was replying, and read aloud.

> As man advances in civilization, and small tribes are united into larger communities, the simplest reason would tell each individual that he ought to extend his social instincts and sympathies to all the members of the same nation, though personally unknown to him. This point being reached, there is only an artificial barrier to prevent his sympathies extending to men of all nations and races.

"That is not a message," Martin added, "that men should tolerate the suffering of anyone."

Will shook his head.

"But that's not the passage men remember, is it? They remember the tales of human progress that root civilization in private property, and place the Indian lower on the ladder, by nature, because he doesn't believe in fences. Because those who refuse to fence their land are doomed by the laws of nature to be extirpated by the advance of those who do."

Martin tried to focus on something he could say with conviction.

"And all that depends on thinking of Indians as a separate race, right? That by nature, as a race, they have no instinct to maintain and defend private property, but hold all in common?"

Will conceded the point by nodding, and Martin pressed on. "But this book doesn't support such claims. Darwin brought every test a taxonomist uses to distinguish species to bear on the question of whether it is safe to delineate human races, and by every criterion—mutual fertility, constancy of character, forms graduating into each other—the assumption that there are discernable boundaries between races fails. There's no animal that has been studied more closely, yet naturalists can't agree whether there are four, five, or sixty-three races of human beings. Any taxonomist would tell you that's evidence the boundaries are arbitrary, and that we have no right to give names to objects that we cannot define." Will looked intrigued, which made Martin even more uncomfortable, so he added, "But Dad, I fail to understand why what biology says should matter a whit. You are talking about how human beings should be treated."

"Because the racialists claim biology is on their side, that's why," Will replied firmly. "And you lot hiding away with a bunch of beetle specimens to keep science pure isn't going to change that fact."

Martin stood up and handed his father *The Descent*. "All right. Read chapter 7. If you don't understand it, I'll tell you how to read it, based on my years of hiding away with a bunch of beetle specimens."

He was tired, even irritated. Losing himself in his scientific work had always been challenging after his father's political lectures. For years having the model of Sam Henshaw sitting in an office down the hall had helped. Out west Trevor Kincaid's consistent shrug in the face of political debates helped justify his own retreat to his microscope when arguments broke out. But it was taking him longer and longer to concentrate on his specimens and forget the demands that every science stand and deliver something useful to American life.

Part III

Revelations

Chapter 9

He could have blamed his dad. Or even Josiah. But having agreed to read up on eugenics before giving James Slater a definite answer, Martin was by his own volition deep in the work of biologists who applied Darwin's theory to human beings.

The only books in his own library that at first sight seemed relevant were a few volumes by the radical physician Havelock Ellis. Will had given the books to him when he left for college.

"Best you not be guided by a bunch of Victorian nonsense when courting your girl of choice," his father had said.

Martin had at first laughed at the gift, and then swallowed his mirth when he realized his father was quite serious. Will had mainly been concerned by the suffering caused by taboos regarding sex, but Ellis's book also included sections on eugenic mate choice. It was in part Martin's vague memory of Ellis's claims that eugenics would alleviate human suffering that had inspired him to consider Slater's proposal with an open mind in the first place.*

Now, a stack of books and articles on eugenics lay on his desk in the Zoology Department. He read the Harvard zoologist George Parker's defense of eugenics as a public service that zoologists were particularly well placed to meet. "It is coming to be a commonplace statement," Parker announced in the pages of *Science*, "that we have paid more attention to the production of high-grade breeds of sheep, cattle, swine, and so forth, than we have to that of effective human beings, and this statement gains popular strength as we awaken one by one to the fact that man is, after all, a member of the animal kingdom and subject to its laws."*

He read Princeton zoologist Edwin Conklin's description of eugenics as zoologists' best means of alleviating suffering, accompanied by severe criticisms of those who took great pride in renouncing "the Devil and all his works, the vain pomp and glory of the world" by keeping to their lab or museum benches. And he read the arguments of the director of the Carnegie Institution's Eugenics Record Office, Charles Davenport, that those who refused to study human problems were stuck in the childlike stage of scientific curiosity. The days when the zoologist, Davenport wrote, "still cherishing in manhood the childish delight of collecting animals and studying with uninhibited enthusiasm the details of their structure," could avoid answering the question "What are your studies good for?" were over. All zoologists must join the effort to collect family pedigrees in order to determine the potential consequences of "racial amalgamations." Only then, he insisted, could that most important element of human conduct, reproduction, be guided scientifically.

This was applied biology with a vengeance, and Martin didn't like any of it. But that fact just made him more suspicious of his own reaction. He suspected his dad's crusades were seeping into his own judgments. He dug through Will's copies of the *Century Magazine* and reread Albert Wiggam's call for a new, biological Ten Commandments, the one his dad had placed before him a few days before they moved to Seattle:

In our day, instead of tablets of stone, burning bushes, prophecies and dreams to reveal His will, He has given men the microscope, the spectroscope, the telescope, the chemist's test tube, and the statistician's curve in order to enable men to make their own revelations. These instruments of divine revelation have not only added an enormous range of new commandments—an entirely new Decalogue—to man's moral codes, but they have supplied him with the technique to put the old ones into effect.

Science, Wiggam wrote, revealed a universe of law instead of a universe of chance, and a God who could be trusted. The Darwinian generalization of organic evolution had finally revealed God's purposes and what was required of the conduct of men and women to ensure salvation. If, he warned, the "monkey legislatures" of states like Kentucky didn't succeed in banning evolution from American schools.*

Faced by the challenge of understanding the visceral rebellion he felt toward all these calls for a new, scientific revelation, Martin was relieved to find, as he and Josiah read part 2 of *The Descent of Man*, that Darwin did not mention human beings for almost four hundred pages. Instead, he presented scores of descriptions of sex differences in insects, fishes, amphibians, reptiles, birds, and, finally, mammals. It took them a long time to get through these chapters. Martin sensed Josiah was also relieved at the change. He peppered Martin with questions about the current status of the inheritance of acquired characteristics and Darwin's theory of sexual selection. It meant Martin had to do his homework between those trips to Tacoma, but also that he could immerse himself in studies of nonhuman animals at least part of the week.

Usually he found Josiah eager to get to work on the specimen list. But one frosty day in late November, Helen informed him as she took his coat and hat that Josiah had had a cold that week and seemed to be taking a longer afternoon nap than usual.

"I'm sorry," Martin said. "I should have called. The collection can wait."

She smiled. "Yes. But he can't. I'm sure he'll be down soon."

Martin entered the study alone, as Helen disappeared to the back of the house. He sat in his usual chair and waited. A stack of trays lay, as always, next to Josiah's chair. He knew Josiah would want to go through them together, so after a few moments he stood and wandered about the room. He went to one of the windows and looked out onto the view of the neighborhood, gray and silent. He could see the base of the tall fir trees he had noticed on his first visit, some in Josiah's yard and the others in the neighbor's. There were dormant flower beds encircling the space between the house and the street. Helen's desk was beside this window. He looked down at the book that lay open, a brass paperweight in the shape of a phoenix on one side. He recognized that book: *The Handbook of Nature Study*, the one his mother had carried to the ponds so long ago. He turned a few of the pages and could almost hear the voice of Anna Comstock, the wife of his own beloved mentor, as he read a few lines:

> Perhaps the most valuable practical lesson the child gets from nature-study is a personal knowledge that nature's laws are not to be evaded. Wherever he looks, he discovers that attempts at such evasion result in suffering and death. A knowledge thus naturally attained of the immutability of nature's 'must' and 'shall not' is in itself a moral education. The realization that the fool as well as the transgressor fares ill in breaking natural laws makes for wisdom in morals as well as in hygiene.*

Though he had lived with the *Handbook* for two decades he had never really noticed these lines, just prior to hundreds of pages of detailed information on plants and animals. Anna's words reflected a firm belief that science provided the means of adjusting one's desires and behavior to natural law. They were a call to look to science as the best means of avoiding suffering. Perhaps even the creation of a heaven on earth.

To his very core, Martin distrusted the ability of anyone to say, precisely, what heaven should be like. He had found no reason to trust eugenists' visions yet. For starters, too much disagreement existed on the knowledge required to get there. Wiggam insisted biologists knew enough now to develop a new Ten Commandments based on evolution. But the Carnegie Institution biologist C. C. Little criticized one of the movement's main spokesmen, Charles Davenport, for doing shoddy work. Little said he'd pay eugenics more mind when its proponents learned to think like experimentalists. Conklin urged that eugenists must pay more attention to the role of the environment, especially given the threat strongly hereditarian statements posed for individuals' belief in the power of reform and personal responsibility. Others warned that surely the laws of heredity were so obscure and complex that no practical difference existed between the working of the laws of heredity and the way in which dice may be taken out of a lucky bag. But all these calls for caution, carefully recorded in Martin's notebook, preserved very little of the claim that biology was useful to humanity now, or that anyone might actually get "nature's 'must' and 'shall not'" right anytime soon.*

He startled, embarrassed, when he realized Helen had noiselessly entered the room with a stack of books in her arms.

"I'm afraid you've caught me snooping," he said by way of apology.

"It's all right," she replied.

He turned the book back to its original page.

"I learned my first natural history lessons from Anna Comstock," he said. "My mother and I used to take this book to a pond and sit with it for hours." He paused, and when she said nothing, added, "You're reading about dragonflies?"

"During our walks Josiah likes to know what is about him," she replied. "There's a pond about a mile from here."

Martin looked down at the book again as she knelt to place the books on a low shelf. He hazarded one more attempt at conversation.

"I hadn't noticed the moral tone of the book until now. Your father has trained me to see that better." He smiled as she looked at him. "Professor Comstock always used to say, 'Be sure you are right. And then look again.' But it's a difficult thing to always remember. To look twice. A third time."

"Yes, professor," she replied with a slight smile in return. "It is."

She stood and went to the specimen cabinet to retrieve a set of trays for them to work on. Martin sat down in his usual chair and pretended to look through one of the trays for a few moments. He wished the gramophone was on, so the music might distract him from an acute awareness of her presence and the questions that suddenly went through his mind. Did she ever have any free time? Did the hours waiting on her father wear on her patience? Did Josiah know of all the ways she moved about the house anticipating his needs? Then Josiah appeared and distracted him with an hour's interrogation of how well Darwin's theory of sexual selection explained a whole range of traits in birds and insects.

Sometimes, Josiah caught Martin off guard with his own questions. They were discussing the chapter on birds when he suddenly asked, "What's it like to be a young bachelor reading these lines?"

Martin's brow furrowed, and he looked up. "I hadn't thought of it that way," he said.

Josiah laughed. "No. I didn't expect you would. Amid all the talk of deer antlers and elephant tusks. Still. If the boundary between man and animal is broken, then all this stuff about the law of battle for the possession of the female appearing to prevail throughout the whole great class of mammals must allow no exceptions. That's your line, isn't it?"

"You're backing me into a pretty inconvenient corner, but yes."

"So our only hope is that men learn to rely on other charms than the ability to win in battle? And that in dealing with his rivals a man is guided by a greater virtue than jealousy?"

"Yes," Martin replied, and then he shrugged. "Of course, Freud says that thinking we can triumph over our bestial instincts by pasting on a bit of civilization is what puts a lot of men and women on the psychoanalyst's couch. Cost of progress, I guess. If he's right."

Josiah gave another, more subdued laugh. "That it is."

Later, Martin couldn't remember whether Helen was in the room when Josiah asked him what it was like to read all this stuff about mate choice as a bachelor. For weeks Martin had become so accustomed to her silence that he forgot to be uncomfortable when he and Josiah discussed *The Descent* as though they were alone. Then they came to the final chapters, where Darwin finally applied his analysis of sexual selection to human beings. Martin had seen some of these lines before. Exhibit designers had placed excerpts at the entrance to the International Congress of Eugenics in New York, and they seemed to occur in some form in every eugenics textbook he'd read over the past few weeks:

Man scans with scrupulous care the character and pedigree of his horses, cattle, and dogs before he matches them; but when he comes to his own marriage he rarely, or never, takes any such care. He is impelled by nearly the same motives as are the lower animals when left to their own free choice.

Darwin had made a few remarks on what was now called eugenics earlier in the book. He had noted that "excepting in the case of man himself, hardly any one is so ignorant as to allow his worst animals to breed." Martin had paused at these passages, because he remembered how Bryan had used them as evidence of the inevitable tendency of Darwinism, even though, as Bryan conceded, Darwin's own heart had rebelled from applying the implicit biological message in practice.

Josiah had made no reference to those earlier passages. He might say something now. For as they reached the final pages of *The Descent*, Darwin proposed that man might, by selection, indeed do something not only for the bodily constitution and frame of his offspring, but for their intellectual and moral qualities. Both sexes, Darwin wrote, ought to refrain from marriage if in any marked degree inferior in body or mind. He conceded such hopes were utopian until the laws of inheritance are thoroughly known, but urged that "all do good service who aid towards this end." Surely Josiah would have something to say given his constant effort to consider the implications of Darwin's work for the age-old endeavor of discerning right from wrong. The passages were good ammunition against Martin's argument that Darwin, as a naturalist, had had very little interest in pontificating about the implications of evolution for human behavior or ethics.

It took Martin a while to diagnose the vague sense of anxiety he felt while reading these chapters. He hoped he wasn't afraid of Josiah besting him in the end. But eventually he figured out what bothered him: his friend could only hear these lines and the chapter about sexual selection in human beings through the calm tones of Helen's voice. Darwin included long, at times risqué descriptions of the grounds on which men and women choose each other as mates. Though he hid some of the choicest passages in Latin, and Martin had no idea whether Helen could understand them, he cringed at the explicit language about males choosing mates based on particular parts of the female anatomy. Having collapsed the boundary between humans and animals, Darwin thought nothing of writing about peacock tales and women's backsides in the same chapter. But Darwin, Martin suspected, had never had to talk about these things with a lady in the room.

Over the space of a few months Helen had read more than six hundred pages of Darwin's words to her father. As he prepared for their final conversation about *The Descent of Man*, Martin's unease regarding the concluding chapters manifested in a wish to find some means of thanking her for the time she had spent ensuring Josiah was ready for their discussions. He scanned the shelves of a bookstore downtown, but he only ever saw her with books Josiah asked her to read, so he had no idea what she might like. And for some reason he could not quite explain, flowers were out of the question.

He finally found something at a little fruit stand in Puyallup, just outside Tacoma. He'd traveled south via the interurban railway a few hours earlier than usual to deliver a set of specimens to the State College of Washington's Experimental Station. An entomologist named Axel Melander, a short man with round glasses and an easy smile, met him at the small train station in Puyallup and drove him the rest of the way. Melander pointed out the sites as they left the flats and entered miles of farmland. Martin happened to ask him about a large building that seemed out of place just above the flats.

"Cushman Indian School," Melander replied, "Or, at least, it was. Got shut down a few years ago. Conditions were apparently pretty bad. Some of the Puyallup who'd been there as kids had organized a public school in Fife years ago, and most of the tribe went there instead anyway. A lot of the kids at Cushman were actually from Alaska."*

Martin remembered his dad's "Alaska, Martin, Alaska!" He turned away from the school and looked straight ahead at the hopfields that spread across the valley before them.

The Experimental Station's insect collection looked very different from the kind Martin had been trained to create. Rather than long series to illustrate geographical variation, these specimen trays contained single adult male and female specimens, a single larva, a single pupa, and examples of host plants. There were probably many things an entomologist could do with this collection, Martin thought as he tried to help Melander with some identifications, but studying evolution wasn't one of them.

As they drove back to the interurban station so Martin could proceed downtown, he asked Melander if they could stop at a farm stand with jars of raspberry jam piled up on a bench. He'd noticed the stand on the drive in, and remembered how, a few weeks ago, Helen had replied to Josiah's request for raspberry jam by saying it was difficult to find raspberries in any form in the markets that season. Yet here, in front of a farm stand run by a Japanese couple, were dozens of jars. Melander obviously knew the husband and wife well, for they exchanged pleasantries about their respective families. As they drove away, Melander told Martin they had owned a farm nearby for decades but lost the land three years ago.

"Alien Land Act barred them from owning land. They sold it to their children, since their kids are American citizens, but Senator Johnson closed that loophole with an amendment. A friend of mine offered to buy the farm and let them continue to work the crops. Some of the families made similar arrangements. The rest have left."*

Melander had provided no political commentary on the facts he shared. But Martin suddenly thought of the passage in *The Descent*, where Darwin had stated with confidence that the simplest reason would tell each individual that he ought to extend his social instincts and sympathies to all nations and races. Melander's account of the couple's history turned those innocent-looking jam jars into a reminder of Darwin's concession that, in fact, experience showed how hard it was for men to obey "simplest reason." Then he thought of how, almost a year ago, Martin had witnessed Erna turn to Trevor after she had told them of the reverential way the Tulalip Indians spoke to the first returning salmon: "Perhaps you just need to speak a bit more nicely to your Japanese oysters."

"Pacific oysters," Trevor corrected her. "Company policy now."

Erna shook her head.

"How do Mr. Tsukimoto and Mr. Miyagi feel about that?"

"Who?" Will had interjected.

"Oyster farmers who came over some years ago from Japan," Trevor explained. "Bought some land on Samish Bay, but all this anti-alien legislation made it very unclear who had title to the land, and they couldn't get loans or sell stock. Company I've been consulting for purchased their land."

Will had let out a loud "humph" as commentary.

"They were going out of business when Mr. Steele offered to buy the company and the land," Trevor added in defense. Then he made a more aggressive stab at his plate than usual. "It was a bad business, I agree." But he was adamant when Erna's brow remained furrowed. "Any biology brought to bear on the question was window dressing, Erna. It was a labor issue, pure and simple. Act would have passed whether the men in Olympia spouted theories about the dangers of miscegenation or not."*

Erna had been willing to let the subject drop, for the children rushed into the dining room all at once with some crisis that needed all Louise's and Erna's attention. But Will, who was hearing all of this for the first time, had let Martin have it on the walk home.

"Kincaid should have refused to have anything to do with the industry. Or the land!"

"I think he just wants to figure out how to make oysters breed, Dad," Martin had replied. "Hard for a naturalist to decline sorting out a mystery like that."

"No doubt," Will sniffed. "Especially when the men in Olympia are willing to pay handsomely for the results."

Now, Martin walked up to Josiah's doorstep with jam jars in hand and alien land acts in his head. Then Helen opened the door and he witnessed her delighted, unguarded smile as he handed her the precious jars. He forgot to explain they were in thanks for the time she had spent reading a rather tedious and indecent (he had not planned to say the latter) book to her father, and entered the study wondering how long it had been since Josiah had witnessed his daughter's gentle smile.

As they talked that afternoon, Martin never learned what Josiah thought of Darwin's description of sexual selection or rational mate choice in humans. For when they set the lady beetles aside and Martin thought they would talk about the final chapters of *The Descent of Man*, Josiah instead felt for something between his chair and the wall and brought out a thinner specimen tray. Martin had not noticed the stack when he arrived. The trays did not belong to the large cabinet that held the lady beetle specimens.

"You know what these are, don't you?" Josiah said quietly as Martin took the tray.

"Ichneumonidae," Martin answered, astonished at the dozens of delicate specimens, their long, wispy ovipositors cast about the tray in a chaotic manner very unlike Josiah's collection of neatly pinned lady beetles.

These were the subject of the poem Pete had read the previous summer, when he had asked whether a man's understanding of God depended on the organisms he studied. Martin had often been impressed by Josiah's tremendous appreciation for the diversity in his collection of lady beetles and by his deep admiration for how Darwin had explained the patterns. He also knew how deeply Josiah struggled to determine the implications of that explanation for human hopes and troubles. Now, faced with two dozen trays of ichneumon wasps amassed over at least four decades, Martin realized that Josiah's effort to wrestle with his beliefs and nature's ways had been long-standing.

He wanted to ask why Josiah hadn't shown these trays to him before. But instead he said, "Do you want us to take these as well? I'm sure Trevor would love to have them. He did some work using ichneumons as biological controls. Before the war." And then he added quietly, as he remembered Pete's words, months ago, about the meaning of these little creatures for belief, "Turns out they might be good for something."

Josiah leaned forward. "Do you know what Darwin thought of these little devils?"

"He cited them as evidence against the existence of a personal, benevolent God. He asked what kind of God would design larvae to hatch inside of caterpillars and consume them alive?"

Josiah sat back in his chair again, and spoke quietly. "I've sometimes wondered whether it's our own fault that these little creatures seem to disprove God's goodness. We're the ones who hung so much of Christian belief on the hinges of a butterfly's wing. And thus, on where the ichneumon place their eggs."

Martin did not know what to say, so he remained silent.

"In training men and women to look to nature for evidence of God's benevolence," Josiah continued, "we turned these hapless, terrible little creatures into damning indictments of faith because they don't fit our own idea of goodness." He took a deep breath, as though he would need it to force the words out. "I have presided over the funerals of children who died because their appendix burst. I have tried to console their mothers, their fathers, amid such loss."

He suddenly stopped speaking. Martin noticed Helen had stepped into the room, and Josiah seemed to hesitate. His mouth opened to speak, then he closed it again. When he finally spoke, it was to recite lines from *The Descent* from memory:

> Man still bears in his bodily frame the indelible stamp of his lowly origin. And because man suffers from the same physical evils with the lower animals, he has no right to expect immunity from the evils consequent on the struggle for existence, subject as he is to natural law.

He paused, and then added firmly, "That is the key, Martin. That is the key."

"The key to what?" Martin asked.

At first Martin thought Josiah would ignore the question, for instead of answering he asked for Martin to retrieve a book from the shelf behind him. "He sometimes cites Winwood Reade's book *The Martyrdom of Man* for information on Africa. There is a passage I would like you to read."

Martin smiled when, after he told Josiah he had found the book, Josiah said the page he wanted was dog-eared. Almost every page in Reade's volume was dog-eared. But eventually, under Josiah's guidance regarding chapter and rough placement, he found the passage requested and read aloud:

> But it is when we open the Book of Nature, that book inscribed in blood and tears; it is when we study the laws regulating life, the laws of productive development, that we see plainly how illusive is this theory that God is Love. In all things there is cruel, profligate, and abandoned waste. Of all the animals that are born a few only can survive; and it is owing to this law that development takes place. The law of Murder is the law of Growth. Life is one long tragedy; creation is one great crime. And not only is there waste in animal and human life, there is also waste in moral life. The instinct of love is planted in the human breast, and that which to some is a solace is to others a torture. How many hearts yearning for affection are blighted in solitude and coldness! How many women seated by their lonely firesides are musing of the days that might have been! How many eyes when they meet these words which remind them of their sorrows will be filled with tears!

Though an unbeliever for whom the passage undermined nothing, Martin stopped reading. Josiah had not signaled to him. Helen had made no movement. He wasn't even sure she was still in the room, and avoided looking up to see. Josiah leaned forward again into the silence of Martin's pause.

"Reade asks whether pain, grief, disease, and death are the inventions of a loving God. He demands that it is useless to say that pain has utility, that massacre has its mercy. For why is it so ordained that bad should be the raw material of good? Some say Darwin triumphed because men and women were tired of being told why an all-wise, benevolent God permitted evil and suffering. That Darwin's explanation of the appendix won his case, not finch beaks or tales about pigeons. For he insisted that our origin proves we must expect nothing more than any other animal who must make its way amid nature's laws."*

"It is a humbling explanation of why we have allies," Martin said when Josiah fell silent. "But you still believe."

"Love, of one's beloved or of God, is not the realm of reason, Martin. Nor, surely, is this search for justice in the system. A man must only decide whether Nature or God, or some collapsed combination of the two, explains the suffering and mystery that surrounds us best." He paused for a long moment. "Of course, what one decides about that question has a great bearing on what, then, one must do. Surely part of the appeal of rooting suffering in nature is that men might learn to control nature, and prevent suffering, rather than submit to it. Even though it provides no purpose to life beyond the hope that our children will suffer less than us."

Martin was on the verge of confessing his dilemma regarding the course on eugenics, for Josiah had somehow captured what worried, even angered, him about the confident statements peppering those textbooks on his desk. In the name of a new salvation, some of his colleagues promised a biological heaven draped in the mantle of science. Just as his lips parted to speak, he heard, or felt, Helen's quiet movement and looked up.

"Shall I serve dinner?" she said.

She was standing at the entrance to the hallway. He had no idea how long she had stood there and feared her timing was a silent signal that he had pressed Josiah too far.

Neither Martin nor Josiah seemed inclined to continue their discussion of Darwin or God over the little banquet she set out before them. Josiah was delighted with the raspberry jam, and after Helen told him Martin had brought it from one of the farm stands in Puyallup, conversation centered on what he'd been doing at the Experimental Station.

"It must be quite a shock to the system," Josiah said, "coming from that kind of work to a collection of lady beetles amassed without any attention to their role in agriculture."

"Yes," Martin said, letting the point pass that they had not, in fact, been looking at the lady beetles. "The men in Olympia are trying to reduce the Experimental Station's appropriations. Melander's had to scare them with threats of locust plagues and the like, should the legislature fail to support its little army of entomologists."

"An army of entomologists to fend off the locusts," Josiah mused.

"And so, the scientists will save us from what once only the prayers of Moses could."

Martin shook his head. "Pretty untrustworthy means of salvation. I can't name one insect outbreak we can actually control. Theoretically one can imagine how it might be done, but there isn't time to amass the knowledge required. Most have turned to chemicals developed during the war. Sam Henshaw got out of applied work because he saw the direction things were going. He was tired of urging farmers and lumbermen to withhold judgment until all the facts were in. Everyone wants productivity saved yesterday. So he went back to the museum."*

"Couldn't serve two mistresses, eh?" Josiah smiled. "You know, they used to say that of the clergy. The main argument for clerical celibacy was that we couldn't be spiritual leaders if entrapped in the petty concerns of the world or wives. Then some realized we couldn't be quite fully human without personal experience of the hopes and troubles of our fellows."

Martin had no reason to think Josiah was talking of him personally. He stared at his plate as Josiah changed the subject back to Melander's studies. He remained distracted as Josiah talked, to the point that when Helen stood to clear the table, without thinking he began stacking his and Josiah's plates.

"It's all right," she said.

"Sorry. Habit," Martin said.

"You either lost your mother at a young age, or she was a feminist," Josiah said.

When Martin smiled and replied "both," Josiah asked him for the first time about his upbringing, and Martin answered as Helen cleared the table. Perhaps the confession of all those ichneumon wasps had taken its toll, though, for Josiah signaled a close to the evening long before Martin must rush to catch his boat. Martin walked him to the bottom of the stairs.

"Full Irish, eh? And an only child?" Josiah said before ascending. "So you and your father are exceptions to Darwin's claim that the careless, squalid, unaspiring Irishman multiplies like rabbits while the frugal, upstanding Scot passes his best years in struggle and celibacy and marries late, leaving few offspring behind him?"

"That was a man named Mr. Greg, not Darwin. He was citing Greg," Martin replied.

"Ah, Martin. It's Darwin. He hides behind another man's words again. Don't you remember? He found consolation for the fear that the degraded members of society increase faster than the virtuous in the fact that the intemperate suffer from a higher rate of mortality and that the death rate in crowded towns during the first five years of life is double that of rural districts."

"I don't think it's fair to say he found consolation in those facts."

"Then your heart is influencing your conclusions." Josiah turned and smiled. "That's not a criticism, by the way." He started up the stairs again. "We will finish the Coccinellidae next week? Now that we can set aside this irritating little distraction of the descent of man!"

Martin went to the parlor to retrieve his satchel and saw Helen kneeling at the fire, spreading the last of the embers out with the poker so they might cool as the house slept. At that moment he realized why he had halted reading Reade's words aloud. It was not because he was afraid of upsetting Josiah. It was Reade's image of women, seated by their lonely firesides musing of days that might have been, as evidence against the theory that God is love, that had made the words catch in his throat.

He knew nothing about her. Nothing about her beliefs or lack of them. Nothing about her past or her regrets. Only that she was obviously quite devoted to Josiah and had an elegant way of writing name after Latin name in a clear, tight script. Josiah's confession during one of his first visits, that her agreement about his character was more important than all their differences, hinted at some underlying struggle between them. It made Martin wonder whether the apparent contentment of this silent, devoted daughter was in fact deceptive. Like the weevils who, when faced with danger, freeze into a position that only the most skillful naturalists and predators recognize as feigning death.

She stood at the sound of his footsteps, wiping her hands on her apron, for she always walked him to the door to hand him his coat and hat. So he once again tried to say what he had meant to say with the jars of raspberry jam.

"I wanted to thank you for the time and trouble you have taken to read *The Descent* to your father."

"It wasn't any trouble," she replied.

He wanted to say, "But it must have been," given that reading the book had taken dozens of hours of her time over the past few months. Instead, he followed her to the door, and replied with a "Goodnight" to her usual "Goodnight, professor."

As the door shut behind him he realized he had never noticed how cool and distant his title could sound.

Sarah used to call him professor, in a light, mocking tone. Once she had used the title with disappointment in her voice. Almost ridicule. It all came rushing back to him as he pulled the door closed and exchanged the warmth of the Grays' home for the cold night air. Some strange mix of Josiah's words after revealing his collection of ichneumon wasps and Helen's farewell combined to open whatever chemical gates had suppressed the memory.

He and Sarah had had an argument. She had been telling him the views of the young minister they had just met at a dinner party. Perhaps she thought those views would be helpful in the wake of his mother's death. Martin remembered now the deep, visceral rebellion he had felt against the minister's belief that individuals must give up the old ideas of prayer. That they must abandon belief in a personal deity and find God's goodness in humanity's increasing ability to understand and control natural law and alleviate suffering.*

She had said something about how "of course" that requires men and women to give up the old miracle stories. The virgin birth of Christ. His divinity. The atonement and resurrection. And thus individual salvation. Sarah had added that she agreed with the minister's view that humanity must abandon its selfish obsession with individual immortality.

She must have thought she was on safe ground. That Martin would have no objections to reliance on science and natural law. But Martin had recoiled at the word "selfish" to describe individuals' hopes and dreams of an afterlife. Sarah must have observed the emotion on his brow. Perhaps that was why she tried to further explain the minister's view that when individuals learned to value the future progress of humanity as a whole, then the death of the individual would matter less, so long as we have lived a life of meaning and done good. God's goodness and design might still be seen in the overall progress of mankind from his ape ancestors to the heights of moral idealism and the League of Nations.

"Your reverend is pretty sure of his ability to say what God is like," Martin replied.

He must have let a tone of sarcasm seep into his words. Or perhaps Sarah had noticed that he did not give his characteristic shrug. But because he had taken refuge in a criticism of the minister rather than the minister's God, Sarah had misinterpreted his anger, and hit another vulnerable mark.

"You're jealous."

Was it with a hopeful tone? Or was his ego misremembering? Maybe it was a tone of disappointment. Or even, perhaps, disdain, as she compared the minister's hopeful alternative to Martin's immobilizing agnosticism.

He had stopped walking and turned to face her.

"No, Sarah. I'm not. I'm sorry. I shouldn't have spoken just now."

She had looked up at him for a moment, before replying, "And when will you speak, professor? Without regret?"

They were no longer talking about whether he was right to criticize the minister. He could see that in her eyes. That was when it had ended. When he said nothing in reply to her demand. Said nothing of what he thought. Of the minister. Of her. Of tomorrow.

He could not understand why the memory came back to haunt him now. That was the trouble with the chemistry of the mind. It was so complicated one might as well believe in free will. And that he had actually known what he was about in letting things with Sarah end.

Chapter 10

Martin often found himself thinking of Pete and Phoebe as he slowly read through the stack of eugenics textbooks on his desk. After his father's, it was the only love affair he had witnessed of late. He knew that Pete was receiving letters from Phoebe via the department mailbox, a pretty good signal he did not want to receive them at home. His young friends' troubles were a constant reminder of the fact parents and elders had long since wished to have a say in the mating choices of America's youth, though the scientific rules being proposed by biologists seemed so different.

James Slater had lent him two texts he used in his own course: Paul Popenoe's *Applied Eugenics* and Herbert Walter's *Genetics: An Introduction to the Study of Heredity*. Popenoe had pretty certain ideas about what students should learn from a course on eugenics. It should, he insisted, inspire faith in salvation via the continuity of the fittest germ plasm rather than individual souls. And it should inspire a heterosexual outlook on life in which actions were chosen based on reason rather than compulsions of an emotional nature. Walter's textbook on genetics urged that biologists should give eugenic advice regarding proposed marriages free of charge. Marriage, he wrote, must be considered from the viewpoint of biology, rather than those of novels, property, status, or sentiment. The fit should be encouraged to marry; the unfit should be told to abstain. Clearly these men would not approve of a smart, attractive, and healthy woman like Phoebe taking a job in a research lab. Or of handsome, young Pete waiting for her.*

Then Martin thought of his dad and Rebecca. One of the books in that stack on his desk was by the Harvard geneticist Edward East, who claimed that miscegenation led to "disharmonious crosses" that destroyed the "harmony of the genotype." Another biologist, Edwin Conklin, critiqued such claims on the grounds hybrids might just as likely be superior as inferior. Neither man questioned whether biology might have something important to say on the matter of who Will and Rebecca should or shouldn't love. And none of them questioned the underlying assumption that the entities they spoke of as "crossing" were, in fact, real.*

Then he learned of another marriage that apparently warranted society's judgment, though he didn't know why or on what grounds. Trevor had just told him that the Board of Trustees had approved the purchase of Josiah's collection thanks in part to Martin's careful inventory when Trevor happened to add: "Did you know Reverend Gray's a bit of a hero among the anthropologists? Erna and Leslie, anyway. I'm not sure why. Spier was surprised he had an insect collection. You haven't met his son, have you?"

The last question knocked any curiosity about what Josiah had to do with Boasian anthropologists out of Martin's mind.

"I didn't know he has a son," he said.

"Must be about your age," Trevor replied. "Erna says they aren't the best of friends."

"Hard to imagine anyone who couldn't call Josiah a friend. Much less his own son."

"Erna says they fell out over his son's marriage."

"That doesn't sound like something Josiah would permit either," Martin said slowly. "The falling out, I mean."

"Well. It's tempting to turn these things into romantic intrigue. Rumor is there was something unorthodox about the marriage. Which, of course, the older generation makes into a demonstration of the end of times, and the young cite as evidence of progress. Way of the world, I suppose. Our bit of it, anyway."

Martin had a hard time returning to his work after that exchange. He had been visiting the Grays' house every Friday evening since early fall. Now it was December. In all that time, Josiah had never mentioned a son, though he had on occasion spoken of the beloved wife who had passed away twenty years ago. The fact Helen hadn't mentioned a brother indicated very little, for Martin had never really had a conversation with her.

On his next visit, he told himself that this little mystery was why his heart beat quicker when, after informing him in her matter-of-fact way that Josiah would be down soon, Helen did not disappear to the back of the house but sat down at her desk. He wanted to ask about the mysterious son Trevor had mentioned. Instead he asked how close the nearest pond was on the grounds Josiah had hoped to take a field trip when the weather warmed. She replied briefly, without looking up from her work.

He gave up. He wouldn't be able to ask her, out of the blue with no warrant or reason, whether she had an estranged brother. Then the realization that his effort to make conversation might be read wrong silenced him completely.

So they went on like this for some weeks more. Until, one evening near Christmas, a knock sounded on the door as Martin and Josiah worked on the ichneumon trays. Helen appeared from the back of the house to answer. Martin heard whispered voices and noticed that Josiah had turned his attention toward the hallway.

"Is something wrong?" Martin said when Josiah began to stand.

"You best go home, Martin. Hand me my cane."

"I'd rather stay. You're troubled by something." He placed the cane in Josiah's hand, and Josiah reached out to place a hand on his arm.

"As my friend, Martin, I would rather you go."

Helen entered the room as Martin reached for his satchel, which was lying on a chair near the doorway. His proximity caused her to look at him before Josiah. His hand froze on the handle. He would defy his friend for a moment more, for Josiah could not see her face as she spoke the words "Josiah, it is Frank, come home."

The tall, dark-haired man who appeared behind Helen was not what Martin expected from his friends' uneasy manner. He entered the room with a pleasant smile and a "Hello Dad" as he removed a tattered workman's cap. Then he turned and noticed Martin's hand still on his satchel.

"Very sorry to interrupt your evening. Put your bag down. You needn't go on my account. Frank Gray," the man said, extending his hand. "You're the zoology professor from Seattle, aren't you?"

"Yes, sir."

He must have betrayed surprise in his voice, for Frank smiled.

"It isn't a difficult fact to determine. I visited an old friend and received a full report on my family. I was informed that a professor joined them every Friday evening. It's Friday and you're here. Sit down, sit down. Please don't stand on my account."

Martin looked at Josiah, but Helen was leading him back to the chair and he was distracted as she whispered something in his ear.

"How do you like the position?" Frank asked. He had walked to Helen's table and was looking down on its contents as he spoke.

"Very much," Martin replied.

"Still. It must be quite a lot of work, covering as much material as you do."

"Yes, it can be."

Frank was at the gramophone now, where he picked out a record from the stack and methodically readied and turned on the device. "Everybody Loves My Baby" drifted from the machine. He then brought the chair from Helen's desk and placed it near the fireside as well. Martin saw his gaze rest on Josiah's copy of *The Descent of Man*, which still lay on the table between them.

"Good Lord, what's this doing here?" he said, taking it in hand. "You're no fundamentalist, Dad, but this is rather shocking."

"We've decided the book isn't quite so shocking after all," Martin said, since Josiah remained silent.

Frank looked at him. "Not shocking?" He smiled slowly. "You aren't reading it right, then."

"Well, we've decided anything in it that's shocking didn't start with Darwin," Martin replied. "That better?"

"Case against God wasn't really sealed until Darwin, though, was it?

German materialists could speak all they liked of thought being to the brain what urine is to the kidneys. They still couldn't explain how the extraordinary mechanisms of either organ had come about without some intelligent force. It left a great deal of room for miracle mongering. Until this." He tossed *The Descent* back on the table. "Church couldn't hold its grip on men's minds if the last rational proof of God's existence lay in tatters."

"Did you plan to stay for dinner?"

It was Helen. She had not left Josiah's side. There was an unease, even a challenge, in her voice.

"Sure, thanks for the invitation," Frank replied and smiled at her.

Martin could not read the glance exchanged, but Helen left the room.

Frank started asking Martin about various faculty at the university. It was a turn to the conversation Martin had not expected. He knew of the McMahons and Vernon Parrington, and asked Martin quite detailed questions about how President Suzzallo was managing his little coterie of radicals. When the gramophone had long since fallen silent and Helen signaled that dinner was on the table, Martin tried to excuse himself again, but Frank protested. This time Martin stood his ground firmly, until Helen spoke to him for the first time since his arrival, asking him to stay as well. There was plenty of food, she added. Martin gave in, because she had never expressed any interest in whether he remained until this moment.

When they were seated, Frank rummaged through a bag at his feet and placed a bottle of contraband on the table. Martin had no idea whether Helen and Josiah were teetotalers or simply law abiding, but no alcohol had ever been served at dinner in his presence here.

"You aren't one of those biologists who think this is some kind of noble experiment to prevent degeneration, are you?" Frank said, holding the flask over a glass he had drawn from the credenza behind him. He obviously knew the layout of the house well. He seemed pleased when Martin let him fill the glass.

"I know some of your colleagues argue against drink on the grounds it subdues reason and gives momentary freedom to vestiges of our animal nature. Might just as well be a useful release, though, eh? No harm in being reminded we have the stuff of warriors in us still. How's that for some applied biology?"*

Martin smiled slightly, but said nothing, and took a swallow of his drink.

"Now. You said that Darwin's book isn't so shocking after all," Frank continued. "Yet the likes of Mencken use it to insist there are lower and higher men, all sorted into their places by Nature rather than God. Surely that shocks something in a man's system?"

Martin panicked a bit at the signal this was going to be the theme of conversation. He knew his H. L. Mencken well enough. He had put up with Ben's parroting of the provocative Baltimore journalist's commentaries on American life for almost two years now. Mencken ridiculed socialists for ignoring nature's law that the strong shall prevail over the weak. If progress depends on the practical enslavement of two-thirds of the human race, Mencken insisted, so be it: nature's laws and human nature could not be cajoled into kindness because men wished for something different. He'd often thought, in the past few weeks, of how Anna Comstock had never pressed her belief in the importance of learning and obeying nature's laws so far.*

"I don't think Mencken is the best source for determining whether Darwin's work is shocking," Martin said. "Or what his work is about at all."

"He's as fine a source as any," Frank replied. "One must take the theory unfettered by an upstanding Victorian gentleman's smoke screen of respectability, and see what men have done with it. Mencken insists science tells us that the statement 'all men are born equal' is sentimentalist nonsense. Says the true message of evolution's that we must simply submit to the status quo and abandon all these grand, impossible plans to change human nature." He looked at Josiah. "Both yours and mine, Dad."

Josiah remained silent.

"Mencken claims to be a great admirer of Professor Huxley," Martin said, trying to find some ground on which he could speak. "But Huxley insisted that our ability to oppose nature, not submit to it, is what makes us human. He didn't believe nature could tell us what we should do and how we should treat others."

"Think men can stand it?" Frank asked. "Left adrift with no compass for determining what is good, true, and right? After all. Huxley took God and the church away as well."

Martin shrugged. "Perhaps not."

Frank was smiling at him. "You don't seem to have much confidence in biologists as guides. I assume you don't want us turning to those who've got religion instead?"

Martin took another drink in the hopes it would stop the vague hints of trembling. The rise of adrenaline in his blood had never happened in this house before. But he still replied at length because by this point he suspected Helen had asked him to stay because he might provide a buffer between father and son.

"I don't know," he said. "If Darwin's right, and religion arises from instinctive impulses, then it seems strange to divide people into those who have religion and those who don't. It might be safer to assume that when a man says he doesn't have religion, he's probably making one. Whether it's science or socialism."

Frank smiled. "And agnosticism?"

Somehow, perhaps from the appeal to Huxley, he had already pegged Martin.

"Sure," Martin replied. "But at least agnosticism acknowledges our frailty."

"But the stance may just be an excuse for biologists to ignore the real meaning of their own doctrines. I've never been able to decide whether that's a heroic or cowardly stance."

Martin smiled. "Neither have I. But I think it's a hard thing to decide what the meaning of any particular biological doctrine might be. It might have no meaning at all, for human beings."

"That doesn't matter if in fact men have found meaning," Frank countered. "The only reason Darwin mattered to Marx was because evolution meant something radical to men and women's vision of themselves."*

"Strange ally, who insists on competition and private property as the source of all progress," Martin said. "I don't think the passages on women in *The Descent of Man* seem very radical to suffragettes. Or those on race very revolutionary to Mahatma Gandhi."

"Everyone picks and chooses the parts they like best," Frank replied. "Kind of like your book of choice, Dad." When Josiah said nothing, Frank continued. "Marx adopted the theory of evolution because it dispensed with miracles, but despised the idea of natural selection. Said the latter's simply a Victorian gentleman reading his own society into nature. The anarchist Kropotkin accepted the parts where Darwin rooted our sympathy and cooperation in instincts and dispensed with competition between individuals entirely. While you lot turned the condition of the poor into the inevitable result of heredity and the struggle for existence."*

"I'm not sure any of that's the work of good biologists, versus politicians and ideologues."

"What the good ones are doing don't matter," Frank replied, "so long as they are silent in the face of the scientific saviors of humanity. Plenty of men are willing to say that from what we've learned about guinea pigs and fruit flies we can produce supermen."

Martin observed that as they talked Frank consumed thrice the alcohol he was. Frank did not interrogate him for long on what some of his colleagues were willing to claim about human beings. Instead, he turned toward Helen, who sat silent at his right. He had not spoken to her, but she must have felt his gaze, for she looked at him.

Frank smiled, and then turned back to Martin, as though nothing had happened.

"And what do you think, Helen?" he spoke, gazing at his glass.

"I do not think . . . ," she started, but Frank turned to her abruptly before she could finish.

"Ah, but you do! That's the whole problem, isn't it? And Josiah sits in silence while you think, and not a word passes between you to sour this sweet Victorian atmosphere!" Frank's voice grew louder. "This is a miserable kind of martyrdom. What rubbish do you have at your desk? Is that anything for a rational creature? You don't think! Hogwash!" He shook his head and pointed at her. "You can't be happy here, Helen. I can see it in your eyes. But you choose this, damn it, and Dad, you stand by and let her pass hour after hour cooking your dinner and darning your damned socks!"

Martin observed much in the next few seconds. Helen's set mouth and furrowed brow hinted at reserves of will that he'd never witnessed when her entire being seemed focused on Josiah's welfare. She stood and tried to move from the table but Frank reached out and grabbed her hand. It was a quick movement, but his confidence in his right to do so conveyed much. Martin realized that quiet, protective Helen was not Frank's sister, but his wife. Though he had been close to intervening to distract Frank from distressing his friends, the knowledge that he must reclassify this woman in his thoughts immobilized him.

"You said you understood this, Frank," she said with a quiet firmness.

"I lied. No one in his right mind could understand it." He turned to Martin. "Do you, professor? Can you understand how a woman who has once lived the life of the mind can abandon it all to spend her days shut up in this house with nothing to do?"

Still gripping Helen's hand, Frank turned his attention to Josiah.

"I don't understand how you can permit this in good conscience, Dad! You can afford to hire a cook and send your socks out for mending!"

But Josiah was a being on whom Martin could get his bearings, and thus defend. He set down his glass with a deliberate movement so that it made a noise hitting the table.

"I hate to break up the party, but I best head to the dock. Are you returning to Seattle tonight?"

Frank looked at him with a stare, as though to demand what right he had to speak now.

"I'd appreciate the company," Martin continued as casually as he could. "Evening boats tend to travel pretty slowly."

Frank had to see through the weak excuse, and for a moment Martin could not tell whether he would take offense or give in. But he held Frank's gaze, and finally Helen's hand was released.

"Capital idea," Frank said as he did so.

"Good. I'll get my coat. Good night, Josiah."

Helen had quickly begun clearing the dishes, so he could not see her face. He was about to say "Good night Miss Gray," but the name was now null and void, a taxonomist's *nomen nudum*, so he said nothing more, and left the room.

Chapter 11

Martin stood in the hall, forgetting that he had said he would get his coat, as he noticed Helen's gray woolen coat hanging beside his. Within a few moments Frank appeared. He seemed sober enough, which made Martin suspect the seemingly drunken outburst had been feigned in the interest of speaking his mind.

"Come on then, get your things."

Martin remembered his coat. They walked for some time without speaking. When they came to the crest of the hill overlooking Commencement Bay, they could see the half moon rising over the mountains, brightening the water. Frank seemed content to take in the view in silence as they descended toward the bay. Though he walked beside a man who might answer the questions pressing on his mind, Martin's dislike of speaking of those not present trapped him in ignorance. They were approaching the fork in the road where one must decide whether to walk down to the bay or turn toward downtown when Frank finally spoke.

"Well, professor, I'd like you to know that despite all appearances, I didn't come here to cause grief to my family. I'm speaking at a meeting downtown at nine. Care to join me? Has something to do with what ol' Mencken is up against. You'll miss all the last boats and trains north, but I have some friends who can put us up for the night."

When he was honest with himself over the next few weeks, Martin had to admit that he said yes because the evening had upset all his classifications of his friends. He might learn something, anything, that would help him re-sort Josiah and Helen more accurately in his mind.

As they walked downtown, Frank began asking questions about his training. He seemed quite interested in Martin's description of his job during the war, his work at Harvard, and of his duties out west. They talked a bit about Josiah's insect cabinets, and Martin noticed Frank spoke respectfully enough about the devotion with which Josiah had built his collection.

"So. What do you and the old man talk about?" Frank asked.

"Lady beetles. Mostly," Martin replied. It was accurate enough since they had finished reading *The Descent of Man*.

"Religion?"

"Sometimes."

"Does she join your discussions?"

"No."

He tried to think of some way to change the subject, but Frank was too quick for him.

"No," he repeated. "She's in the kitchen." After a few more paces he added, "She doesn't like to cook, you know. Never took any interest in it before. Now she cooks three meals a day for him. Know how much time that takes? And the laundering? She doesn't send it out." Then, with undisguised exasperation, he said, "You know, professor, she could tell you both a thing or two."

Martin said nothing.

"For a scientist you aren't very curious," Frank pressed.

"It's not for lack of having questions," Martin replied. "But the answers would be given by a jury of one where three are concerned."

"You're very disciplined in your interrogations. That doesn't come from your science. Biologists build fairy tales on insufficient evidence as well as anyone else. The lives of your friends must be the object of some study. You must have drawn some conclusions."

Martin shook his head. "You are Josiah's son. I know that now. You don't seem to be on the most comfortable of terms. I've heard explanations that must be rejected in the face of evidence. And I am left with no alternative hypothesis."

"And you make none of your own."

"No. I make none of my own. It has nothing to do with me."

"When has that ever stopped inquiry?" Frank exclaimed. "What does a single unnamed insect have to do with you? And here, at your fingertips, under your very nose, are two human beings who are your friends. I'm quite obviously a source of misery to them both. Surely you've concluded that much. Yet you tolerate me as though you've observed nothing."

The dig at Martin's skills at observation touched a nerve, but he spoke with firm calmness.

"Has it occurred to you that I tolerate you because you upset my friends and I wanted you out of their house? You spoke of alcohol as releasing inhibitions. But those inhibitions might be useful. They might prevent us from hurting those we care about. And I can bear to tolerate you. You have no connection to me beyond these moments."

Frank fell silent for two blocks.

"I shouldn't have come, you know," he said finally.

"Then perhaps you should leave."

As the words escaped, Martin realized he must have drunk too much.

"I could box you for that," Frank replied, stopping sharp under the light of a streetlamp. "Don't think I couldn't. And I have the law behind me."

But Martin held his gaze once more, despite the strange insinuation that he was somehow a rival for a woman who barely spoke to him. Then Frank shook his head, as though to escape some thought, and kept walking.

"A man who walks with someone who speaks his conscience with a glance is not his own. Damn it all."

He began rummaging through his satchel as they walked.

"We're almost there. Before I forget, I meant to leave this for Helen. Mind giving it to her? Probably doesn't have time for it now she's taken up cooking. But I promised."

Martin took the book. They were between streetlamps so he could not see the title, but the volume was thick and heavy. He tucked it into his own satchel and glanced at some figures in an alleyway as another streetlamp flickered in front of them. They were not in a respectable part of town.

"So where are we going?" he asked.

"Almost there," Frank replied.

Frank guided him to a dilapidated storefront near the railroad station, a low-rent district home mainly to missions and brothels. The cheap saloons had been closed down by prohibition. Some remained empty and abandoned, and others were graced by hastily created placards advertising soda and sandwiches. Frank stopped at a shop with IWW painted in large, red letters on the door. Martin knew there must be one of these offices somewhere in Tacoma. His dad had told him, with a delighted laugh, that the Cushman Indian School couldn't hold its teachers very long because Tacoma's Wobblies recruited them so quickly.*

Faded Industrial Workers of the World pamphlets and posters were on display under the streetlamp. Martin knew the look of those pamphlets well. Dozens were piled up in one of his father's Indian baskets, worrisome evidence of where Will was spending at least some evenings. But now it was Martin who stood on the threshold of a Wobbly office.

Two large young men barricaded them from entering before Frank said, "Easy there, comrades. I'm the guest speaker for the evening." He held out his hand. "Frank Gray."

The men's demeanor changed from a menacing barrier to an enthusiastic welcome as they shook hands and stepped aside with a "Sorry, doc!"

"They're a bit antsy about raids," Frank explained once they were inside. "Poor lads think the law's still paying attention."

Martin was soon forgotten as the men within crowded around Frank. Backing out of the circle, he bumped up against a bookshelf near a small woodstove whose embers were dying out. He considered adding some wood to the fire but instead blew on his fingers and watched as Frank gestured with animation in response to questions from his little crowd. Eventually Martin grew tired of trying to lip-read, and turned to browse the battered, dog-eared copies of Darwin, Herbert Spencer, Voltaire, and

Thomas Paine. When Frank found him again with a "Hey, professor," Martin had immersed himself in the pages of Paine's *Common Sense*.*

One of the more well-dressed men in the room stood at Frank's side.

"This is Harry Ault. Editor of the *Seattle Union Record*. He's skipping out now. Probably afraid he'll have a conversion experience during my speech. He tends to keep to more respectable lines of work."

Harry smiled. "I'd love to have my enemies hear you say so."

Frank shook his head. "We thought we were going to have a real revolution here," he said to Martin. "After the war. Now the work's been trumped by compromises and complacency. I don't mind saying I hold all your negotiating with industrialists responsible, Harry."*

"Well. Each individual must at some point make a decision, how to best carry on the fight," Harry replied. "It's pretty unlikely we'll all decide on the same strategies."

He turned to Martin. "Frank tells me you're a biologist."

"Entomologist, yes."

"What do you think of Congress using this biological stuff to justify shutting the door to immigrants from Eastern and Southern Europe? Got a bunch of biologists to testify that immigrants can't understand American principles of government because their politics are rooted in their germ plasm."

"A few biologists," Martin corrected. "Charles Davenport. Harry Laughlin. They don't stand in for all of us."

"One wouldn't know that from their testimony or the newspapers. *New York Times* just printed some biology prof from Virginia's address on 'What Biology Says to the Man of Today.' Says restrictive immigration laws will maintain the purity of the white race, and that the myth of the melting pot is bungling human evolution. Stanford's Professor Jordan testified in favor of the act as well. Insisted it's a scientific fact America's best blood's being diluted by an influx of peoples who are biologically incapable of rising above the mentality of a twelve-year-old child."*

"There were also biologists who testified against those claims," Martin said.

He had had to follow the hearings closely because Will had been home for the weeks in March 1924 when they were all over the newspapers. So he could thank his dad for the fact he knew Herbert Jennings had stood before the House committee and pointed out that if one really wants to divide Europe into desirables and undesirables, it was a northern country that had the most insanity, crime, and feeblemindedness. But Harry Ault had paid attention too.*

"Didn't do any good," he said. "They still passed the thing. And Jennings's main concern was that the act didn't bar the Irish! These men are cloaking race prejudice, pure and simple, with words like dolichocephalic, brachycephalic, Nordic, miscegenation and the biological viewpoint, rather than helping us reason ourselves out of our tribal hatred for our fellows."*

"I don't think there's a biological viewpoint on these matters," Martin replied firmly. "Not one that can be trusted."

Harry nodded. "Glad to hear it. But I don't think that's common knowledge." He paused. "I understand why you're irritated by my questions then. The legislature in Olympia asked me to give the labor viewpoint when they were debating the Alien Land Act. Imagine that. Decades of men and women fighting over what it means to be part of the labor movement, and they ask me to give the labor viewpoint."

"What did you say?" Frank asked.

"I said my stance on immigration is based on economics, not racial prejudice. You should have seen the chairman's face: 'How's that?!' he spluttered. I said I didn't think the color of a man's skin was a fair estimate of his right to life, liberty, and the pursuit of happiness. All they could think about was whether I extended that view to the right of any man to marry the millionaire's daughter across the street. I told them I thought such a question none of our business. That baffled the lot. Demanded to know whether I applied that line of thinking to marriages between Africans and Europeans. It wasn't any use to throw Booker T. Washington or Frederick Douglass at them. They whined about the laws of nature. And I ask you, on what grounds can you meet a man when he differs on the laws of nature?"*

"Surely men denounce race equality on the grounds God created the races separate just as often," Martin said.*

Harry nodded. "Oh, sure. But everyone wants to look scientific nowadays. No doubt it would have been more honest for the senator to have said 'We hate the Japanese and therefore they should be excluded' instead of making assertions about either God or Nature. Irony is that the ministers stepped up and opposed the Alien Land Act loudest. Old Mark Matthews, God Bless him. For once."*

"I was afraid you're getting a bit too chummy with such company," Frank said.

Harry just shook his head and smiled as someone rapped a podium at the front of the room and he put on his hat. "Well, I best catch my boat before I'm corrupted." He reached out to shake Frank's hand, and then tipped his hat in Martin's direction. "Keep your scientific cap on, professor. He's pretty persuasive."

The two dozen or so men, and a few women, present sat down in rickety seats and someone stood to introduce the speaker. Within moments, a whole series of facts were placed before Martin that, together, told at least part of the story of Frank's departure, absence, and unexpected homecoming. Martin learned that the "doc" was earned in medical school, but that Frank had abandoned medicine when he realized he was patching up the bodies of men, women, and children only to send them out to be destroyed by the crushing wedges of capitalism. He had traveled to Russia to witness the results of the Bolshevik Revolution. Now he had come home. There was, of course, no mention of an estranged wife or a blind father left behind. Martin knew from years of his father's manifestos that those facts would have seemed superfluous to the battle being waged to free the proletariat from oppression.

Frank stood to eager applause as he was invited to the front of the room to speak. He began by praising the local socialist newspaper the *Northwest Worker* for bravely explaining the true origins of the war: "Start a war with Germany, no matter what for. This will settle the employment problem for some time." Somehow, Frank urged, men must learn to see the cries of "patriotism" that had led millions of workers to a meaningless death for what they are: demands to fatten Rockefeller, Morgan, Carnegie, Rothschild, Guggenheim, and other industrial pirates' bloodstained coffers.*

"We took our stand: rebellion sooner than war, and they imprisoned many of our comrades as traitors," he said slowly. "So be it, if a man who insists he belongs to no country, in the interest of the brotherhood of man, is a traitor, he is a traitor to a system from which the heart and the mind rebel."

Then Frank took on the ideas, as though he were lecturing a hundred men of influence, rather than the transients and day laborers before him. He told of men who believed that Nature or God placed each man in a station assigned him. He ridiculed the timber barons in the region, the Weyerhaeusers and the owners of the Northern Pacific Railroad, for their belief that men and women were mired in poverty by decree of heredity or providence, rather than an unjust economic system. Then he suggested an experiment, to settle the matter once and for all.

"I say get rid of the toil that blunts children's brains and the disease that weakens their bodies. Give them the same nourishment and amount of sunlight as their betters and see whether they are the victims of congenital weakness! Feed the mothers better and observe whether they have stronger babies! Try that experiment! And until you do, don't tell us that poverty and disease are inborn as an excuse for doing nothing!"

He looked at Martin, who was leaning forward with his elbows on his knees, in the back, listening.

"But it's not just our enemies who are mired in visions of what they think Nature tells them. We too undermine the revolution when we find hope in some inevitable evolutionary path toward progress. The men and women at the Seattle Labor College call themselves Christian socialists, place God behind evolution, and trust natural law will damn the bosses eventually. If John P. Weyerhaeuser had any brains, he'd throw a few million dollars at the college and honor it for encouraging complacency. Men who call themselves socialists are using the laws of nature as an excuse to abandon the fight on the grounds wage slavery and private property are parts of some divine scheme drawing us toward the blessed time."

He looked from one side of the room to the other as he spoke, as though every individual present might need the reminder of what they stood for.

"Comrades, to accept this evolutionary fairy tale is to stop all effort. It deadens action and sets human beings against each other! I've heard socialists justify the segregation of the man whose ancestors are African on the ground he is by nature at a lesser stage of civilization. I know socialists who justify the extermination of the Indian because evolution tells us his fate is the biological price of progress. Who speak against the Japanese on the grounds it is a natural law that the race struggle trumps the class struggle. Who say it is natural to feel these boundaries and that race hatred is a biological, protective barrier against amalgamation! We must see this rubbish for what it is! Teaching race hatred is the foundation on which capitalists divide workers. It is how men were inspired to enlist in the bloodiest war in history. It is how workers are pitted against one another, and it is what will destroy the revolution if we stand by and do nothing! We must have one big union, that a man may join no matter his color, race, or creed! This is the stance of the Industrial Workers of the World! It must be the stance!"*

Frank's concluding flourish was met with passionate applause. Someone thought the moment right to strike up a Wobbly hymn. It was sung to the tune of "Bye and Bye" and, Martin knew, one of the milder songs in the songbook:

> Long-haired preachers come out every night
> Try to tell you what's wrong and what's right
> But when asked how 'bout something to eat
> They will answer with voices so sweet:

> You will eat, bye and bye,
> In that glorious land above the sky.
> Work and pray, live on hay,
> You'll get pie in the sky when you die.

The starvation army they play,
They sing and they clap and they pray.
Till they get all your coin on the drum,
Then they tell you when you are on the bum:

> You will eat, bye and bye,
> In that glorious land above the sky.
> Work and pray, live on hay,
> You'll get pie in the sky when you die.

Holly Rollers and jumpers come out,
They holler, they jump and they shout.
Give your money to Jesus they say,
He will cure all diseases today.

If you fight hard for children and wife—
Try to get something good in this life—
You're a sinner and bad man, they tell,
When you die you will sure go to hell.

Workingmen of all countries unite,
Side by side we for freedom will fight:
When the world and its wealth we have gained
To the grafters we'll sing this refrain:

> You will eat, bye and bye,
> When you've learned how to cook and to fry.
> Chop some wood, 'twill do you good,
> And you'll eat in the sweet bye and bye.

The obligate singing of "Solidarity" followed. As they sang, Martin was surprised to catch the eye of a young man sitting in the back row who looked familiar. The man touched his hat to acknowledge Martin's gaze.

Before Martin could place a name to the face, the door near him burst open and police streamed into the room. One of the policemen grabbed a man near Martin by the collar, and without thinking Martin jumped up and shouted "Hey!" Another policeman near him shouted something incoherent, and then everything went blank.

—

It all came back very slowly. The memory of Frank hollering something about the police having the blood of an innocent man on their hands. Of trying to focus on alternating lines of light and dark and suddenly realizing they were formed by the bars of a jail cell. Of retching at the smell of vomit, and realizing it was his own, all over his shirt. Of Frank leaning over him and speaking with urgent firmness: "Listen, professor, you've got a bit of a concussion. I've made it sound worse, well, quite fatal in fact, to get you out of here. You're going to be fine. Tell Helen I said you're going to be fine." Of a policeman on one side and Frank on the other, helping him to stand. The apparition of Helen and a strange man appearing in front of him. Her voice repeating "His satchel, he had a satchel," until a policeman pressed it to his chest and said something about "staying out of trouble in future." Somehow, he got in a car, and somehow, he found himself sitting in his usual chair at the Grays' home.

Only it wasn't Josiah in front of him, sitting back and seeing nothing. It was Helen, kneeling on the rug and steadying his head with one hand, forcing him to look her in the eyes. She asked him his name, and he answered so firmly, to alleviate the anxiety in her gaze, that he winced at what speaking did to the pain in his head. Then she asked if he knew who she was. He said nothing. She misunderstood the source of his confused gaze, and there was a hushed discussion of whether to call a doctor. He remembered the sound of his own voice repeating what Frank had said, and that Frank had said it. The man helping him out of his shirt, and Helen giving orders where to find one of Josiah's in the pile of laundry in the kitchen. Someone bandaging his head. A fitful sleep on the sofa. Helen kneeling beside him, a gentle hand on his shoulder to wake him before sunrise. Her quiet voice asking where he lived, because they must get him home. The man coaxing him to stand and placing an arm around his waist to help him walk. Helen standing in the doorway with so weary a look that he mumbled an apology for all the trouble he had caused.

For half the drive he lay in the back of the car, coming in and out of consciousness. When the man, who he now knew was Charles because Helen had said his name, pulled up to Mrs. Macleod's house, Martin had never been so thankful that his dad was on another week-long adventure up north. He was able to convince Charles that he could manage from the curb just fine. Upstairs, he set his satchel down, lay down on his bed, and stared at the ceiling as his head pounded.

Frank had admonished him for not asking any questions. Now they pressed on his mind mercilessly, as confusion gradually gave way to the longing to get his knowledge of the beings around him accurate. As an expert in determining natural relationships he was astonished by how deeply his powers had failed him. Despite all his training to look closer, distrust casual observation, and be wary of appearances. Despite the pride he took in not being fooled by false resemblance. He had classified Helen wrong.

He thought of the confrontation at dinner and Frank's insistence, during their walk downtown, that she could tell them a thing or two. He thought of her silence, and of the fact he must place her, by marriage, within a circle of hounded radicals. A saying of the great French naturalist Cuvier came into his throbbing head: "In order to name well, you must know well." It was a hard thing to accept, as he tried to adjust the circles he had drawn around Josiah and Helen, this inability to gather knowledge about a fellow human being's past, though they were so near. Especially when his mother had taught him that misdiagnosis was not only a matter of ignorance but of justice.

Chapter 12

"Christ, Marty! What happened?"

Martin woke up to find his dad and Rebecca leaning over him. Will held a bunch of bloody bandages in his hand. They had loosened as Martin slept, and by morning the wound on his forehead had opened and half his pillow was bright red. Martin tried to sit up, but Rebecca placed a hand on his shoulder.

"Wait," Will said firmly.

Martin could hear him rushing about the apartment. Rebecca sat on the side of the bed, her hand still on his shoulder, as though she didn't quite trust him to obey his father. Martin closed his eyes, trying to concentrate on the weight of her hand rather than the pain in his head.

When he opened his eyes again he saw his father kneeling down, a bowl in his hands as Rebecca leaned forward and gently cleaned his forehead. The iodine stung sharply and he jerked back. Rebecca laughed at him, and he couldn't help but laugh as well, though it hurt to do so.

"Now, be still, professor," she said to Martin. To Will, who was watching her anxiously, she said, "The cut's superficial, but foreheads bleed a lot."

It was only after she had placed an adhesive bandage over the cut, and Martin proved he could stand on his own by walking to their breakfast table, that Will asked his question again. Martin rested his head in his hands as he answered.

"I got caught up in a raid on a Wobbly meeting. In Tacoma."

Will stared at him. "What the hell were you doing with them?"

Martin shrugged, though the movement made the pain in his head worse. "Research?"

His dad was plainly unsatisfied with his flippant reply and waited in silence for more. So Martin tried to explain.

"Josiah's son showed up last night. He was speaking at the meeting. Just got back from Russia."

"No kidding?" Will said under his breath, and then he slowly smiled. "Don't tell me I need to watch the company you keep?"

"No. I'll be sticking to my desk from now on." Martin looked at Rebecca. "Thank you."

She pressed his hand gently. "Mind you take it easy for a few days."

"I'll make sure of that." Will said.

Will stayed home for a week, the longest he had been in their apartment since he had met Erna Gunther more than a year ago. He refused to let Martin read anything. Once his eyes could stand the light again Martin spent much of the time bundled up in blankets on the deck, watching the winter sky. Until he couldn't stand the stillness any longer and implored permission for a walk.

"Sure, Marty. Where do you want to go?"

"Anywhere."

For three blocks Will insisted on holding Martin's arm, until Martin said something about how he liked Rebecca.

"Notice how nice she did that bandage?" Will replied with obvious delight. "She's almost finished a nursing course."

Will's talk passed back and forth between singing Rebecca's praises and criticizing the Bureau of Indian Affairs for several blocks, and he soon forgot his patient or how far they had walked.

"There's one thing I don't understand about her, though," he said at one point. "She goes to this Shaker church on the reservation."

"What church?"

"Shaker. Not the English kind. It's an Indian church. Started by some mystic from an island near Tacoma. I think you'd lost consciousness when she said some incantations over you. At least, I think that's what they were. She's got a pretty voice for them. Makes up for the nonsense."

Martin was a bit astonished. His dad was obviously bewildered by Rebecca's beliefs, but once more a humility accompanied his words that Martin had never heard before when his dad talked about any religion. "It's like the words catch you up, though I couldn't even understand them."

As usual, he found something to get indignant about pretty quickly.

"All against the rules, you know. Catholic sermons on eternal damnation. Protestant sermons on original sin. That's all fine and respectable. But an Indian who's discovered Jesus on his own terms? That's going a damned sight too far. Couple of the Tulalip were put in jail for attending Shaker services. Indian agent finally let them out when someone pointed out the Constitution guarantees freedom of religion. No exceptions."*

By the time Will remembered Martin might not be well enough for a long walk yet, they had walked four miles from the house and Martin was starting to feel like himself again.

—

Going through his satchel a few days later as he got ready to go to work, Martin discovered the book Frank had asked him to give to Helen. He turned the thick red volume over in his hand and read the gold-lettered print on the binding: *Medieval English Nunneries*, by Eileen Power. Written on the title page were the words *For Helen. Affectionately, E. E. Power*. A paper fell out of the book's pages and onto the floor. His headaches had lessened, but when he bent down to retrieve it he winced as the movement sent pain up his temples again. He had to pause and sit down on the bed before unfolding the paper. *Vassar College, Poughkeepsie, New York* headed a clear, slanted script. Having identified the contents as personal correspondence from a glance, he knew he should have folded it again and placed it back in the book. But he kept reading, telling himself the next line would be the last even as he ignored the command.

November 25, 1924

My Dearest Helen,

I have enclosed Eileen's book, which she asked me to send to you. It has been sitting on my desk for some time, for I knew I must send an apology in the parcel as well. I am afraid that when we parted so suddenly I could think of little else than my own loss. With that man standing beside you, I could not speak the reasons you must stay. My protest against your leaving us would have seemed inseparable from the fact of your marriage. Now he has written that he no longer has any hold on you. I know he tells me this out of no concern for my own wishes or our profession. I must believe that he has your interests and wishes at heart in telling me you are now bound to no one.

Now that I can speak freely, if I cannot help writing words that appear unfeeling, it is because I feel so much at the loss of you. Let me be absolutely honest, out of respect for you and out of respect for your scruples. Though the word implies disdain, I mean no insult. I take it we are dealing with your conscience. I never attributed your departure to your marriage. Those who do not know you well no doubt saw your leaving as the inevitable result of finding a husband. I know your marriage had nothing to do with your decision, apart, perhaps, from the ideology to which your husband is linked. You know the arguments I would bring to bear against the claim that our field has no utility but to strengthen the hand of the victors and the powers that be. History can be a far more radical thing than Frank allows. Need I add that working on something in which you can put your entire mind and heart, as you once could do, can also be of some solace amid great loss? When work is all you have, at least you have your work.

I am not urging a duty to return to us upon you as some distant ideal. Crusades count for little when one has no prospect of earning one's bread. I have a concrete proposal. I wish to put your name forward for the professorship in medieval history that is opening due to my retirement. Please write and tell me whether my nomination would be in accordance with your wishes. I hope that it is.

Yours very truly,

Lucy

PS. You know I do not care about the particular arrangements between you and Frank. I cannot guarantee that the College's trustees will
be so indifferent. Thus, you should be prepared for the potential of some
inquiry into the subject. As you know, a woman's allegiance to convention is often more important than her politics. But I do, from the bottom
of my heart, urge you to bear up against that inconvenience for the sake
of the good you could do here.

He had to return the letter to her. But something in the lines made
his heart race at the thought of confessing that he had read the entire
thing. There was nothing incriminating in the correspondence. But he
feared that to hand it to her with the words "I read this" would amount to
a challenge, like Frank's, of the fact she had abandoned her intellectual life.
Then, when he momentarily stopped worrying about how to face Helen,
he wondered why Frank had given him the book, when he could have
handed it to her easily enough.

Frank had ridiculed him for not asking questions, but now, with this
letter in hand, a dozen more pressed on his mind. He got through his
classes and a department meeting in a bit of a daze, trying to ignore the
stares at the bandage on his forehead. By the evening he was exhausted
and his head hurt but instead of going home he went to the library. He
retrieved the *Educational Directory* for 1924 from the reference desk. He
usually used this book to locate curators of university museums. Now he
thumbed to the entry on Vassar College, and, scanning the names, found
a Professor Lucy Maynard Salmon listed as a faculty member in the History Department. There were a few books under her name in the card catalog, and he soon sat on the floor in the stacks, trying to reclassify Helen
by delving into these vague clues about her past.

He read several pages of Professor Salmon's books, each consisting of
detailed analyses of the challenges of studying the past. Nothing escaped

Salmon's attention as possible sources for reconstructing the life of individuals and societies long since gone. Cookbooks and abbey inventories in the Middle Ages. Newspaper advertisements more recently. Something in the effort reminded Martin of Darwin's attempts to reconstruct the connection between living three-toed sloths and extinct giant ground sloths, with nothing to come to his aid but a stack of silent bones. They were both fascinating puzzles created by silences in the historical record, problems Martin had dealt with for years, albeit in so different a field, as he tried to piece together the evolutionary relationships of beetles.*

He scrambled up from his pile of books when the lights went out and reached the doors just in time for an annoyed watchman to let him out. New facts tumbled through his head as he walked home. Professor Salmon wrote of the many elements of uncertainty that enter into a woman's life that do not hamper men. Women learned all kinds of skills like sewing and cooking but nothing of the universe of truth beyond such things. Yet Helen had entered that universe and, according to Professor Salmon, had loved it.

With a sinking feeling, Martin remembered all the things Helen had read in order to prepare Josiah for their discussions of *The Descent of Man*: that women's maternal instincts made her more tender and less selfish, while men were competitive, courageous, pugnacious, and energetic, with a more "inventive genius." That "if two lists were made of the most eminent men and women in poetry, painting, sculpture, music, history, science, and philosophy, with half-a-dozen names under each subject, the two lists would not bear comparison." Martin had smiled when he first read these claims, for he imagined Phoebe reciting them in the voice of a high-class British gentleman. He was ashamed that he had stopped wondering what Helen thought, simply because she was silent.

Late that evening, he picked up the book Frank had given him once more. He lay down on his bed and began reading, as though the pages might contain clues regarding her thoughts and cares. He eventually fell asleep with random facts about nunneries in the twelfth century in his head.

The fatigue and vague throbbing from the concussion still tended to set in by each afternoon, so when he returned home from campus Tuesday evening, he nodded absently when his dad met him in the hall and informed him there was a guest waiting for him in Mrs. Macleod's sitting room.

"She's been waiting for some time, my boy. Wouldn't just go to the university and find your office as I suggested."

Martin looked up at the "she." He set his satchel down and went into the sitting room.

"Helen," he murmured in surprise, though he had never used her first name before.

"I'm sorry for bothering you at home, professor."

She had placed her coat and hat on the sofa but was holding a brown paper package. Mrs. Macleod had one of the new radiators and kept it on full blast throughout the winter. Martin wished he could take off his own jacket, but the familiarity implied by doing so prevented him, which meant he assumed the source of the sudden warmth in his frame was the radiator in the corner.

"It's no bother," he said. "Please sit down."

She ignored the request. She was looking at his forehead. He guessed at what the concern on her brow meant. And why she was here.

"I'm all right. Bit of a headache when I stand up too quickly." He smiled. "So I stand up slowly."

She did not return the smile. "I'm sorry I had to ask Charles to drive you home before daybreak. I didn't want Josiah to know how much trouble Frank got you in."

"It's all right. I understand. I appreciate your getting me home."

He meant to add that he was glad she had come, for he could return her book and letter, but suddenly he was thinking of this man Charles. He fell silent, bewildered by the sudden vague sense of anxiety that followed her speaking the man's first name.

She held out the package. "I brought your shirt."

"I'm afraid I can't get the stain out of the one you lent me," he said, taking the package in hand.

"He has plenty of others." She hesitated before speaking again, her brow furrowed. "I'm so sorry about everything. You might have been seriously hurt."

"It wasn't your fault," Martin said.

Given he had so rarely heard her voice, much less directed at him, he was astonished by the candor of her reply.

"You had to intervene because I didn't. I know what Frank is like when he drinks. I know how to calm him down and get his sights off Josiah. If I had done what I should have, you wouldn't have had to escort him out. But I was ashamed that you had to learn by his showing up at our door what we did not tell you. I came here because I can't easily say this in front of Josiah. It would pain him too much."

Martin frowned. This evidence that she cared what he thought about her did not line up with how she had dealt with him for months. But he tried to match her candor by his own confession.

"I must admit I was surprised Josiah has a son. He was never mentioned. Though I believe I may be counted as a friend."

She nodded. "I was afraid you would feel deceived."

"No. Not deceived. I drew hasty conclusions, and that's my own fault. I looked no further than appearances. I'm sorry for that."

"You had no reason to classify us any differently."

She smiled ever so slightly as she referred to the skill on which he had based a career. She must have tried to read his face.

"You're angry," she said.

"No, no. Just . . . confused." Something in the sympathy with which she tried to diagnose his feelings inspired Martin to speak the question that had been in his thoughts since Frank's visit, but that he had never expected to actually voice. "Can I ask you something? Why have you been silent, all this time, as though you had no interest in anything we talked about?" She said nothing, so he added, "Frank told me you could tell us a thing or two."

With another slight smile, she replied, "Frank likes to see a great deal of himself in other people."

That deflective reply let Martin's frustration with himself and the situation escape.

"But since his visit I've been imagining that all sorts of silent, humanist barbs have been coming from your corner whenever I said anything foolish."

"You're making quite a few assumptions about what I might think."

"And you don't care to correct me."

"No. I trust you to remember your ignorance soon enough. You always do in your conversations with Josiah."

"Then you do listen."

"Of course I do."

She turned toward the sofa and picked up her coat. He couldn't tell whether she was annoyed, hurt, or indifferent to his question and confession.

"Wait just a moment," he said finally. "Frank gave me something that belongs to you."

He rushed up to his own room, took off his jacket without thinking, and retrieved the book. When he faced her again and handed the items to her, he spoke before she even knew the letter's contents.

"I'm afraid I read the letter. It was in the book and fell out. I'm sorry. I should have folded it back up."

She glanced up at him briefly, and then, laying her things aside once more, opened the letter as she sat down on the sofa. For some moments Martin was forgotten. He sat down too, in a chair across from her, trying not to betray curiosity regarding her response to this appeal from her past. He kept his eyes on her shoes. They were plain, practical, and flat-footed. He thought of a clipping from the *Century Magazine* that his father had read aloud one evening when arguing with Sarah. It was a debate between Charlotte Perkins Gilman and some man who insisted women wore high heels because they liked them. "Nonsense!" Gilman had replied: women had been brought up to obey the desires of men rather than their own comfort, and men thought the feminine foot must be small, feeble, and slenderly curved. She offered as proof "patent and pathetic" of her theory that those women who rebel against masculine demands and do not "doll up" are "let alone." They are not danced with, nor invited about.*

Had she ever danced much, or, before Frank, been "invited about"? Did this man Charles, who had driven him back to Seattle, ever take her dancing? She finally finished and folded the letter. Martin spoke quickly.

"Frank must have forgotten to give it to you when he had the chance." She nodded and placed the letter back in the book.

"He does tend to forget little things like this." She looked up at him, and he could see her looking at his forehead again. "You're going to have a scar."

He smiled. She had changed the subject. "Cost of curiosity," he said.

She smiled in return, but it was not the open smile that had stopped him in his tracks when he had handed her the jars of raspberry jam. She picked up her coat as a preface to standing again. Martin justified his response to that movement by the fact the journey from Tacoma had taken hours and by his dad's report she had waited for him for some time.

"Wait. You've come all this way. Let me at least buy you dinner. There's a café just a few blocks from here."

"Thank you, professor. But I best get back."

Her reference to the obligations that determined her moments, and her refusal of his invitation, made him speak the questions that had pummeled his mind for days.

"Is it because of Josiah? Is he why you don't join our conversations?"

"No. Josiah is not the reason," she replied.

Martin pressed on, because this might be his last chance. "Can I be completely candid with you, Miss . . ." He halted. "What should I call you?"

"Helen," she said, as though that were obvious.

"Helen, then." He took a deep breath. "Don't you think it has been a bit unjust, sitting near, all this time, silent, as though you had no interest in anything Josiah and I talked about?" He struggled to explain himself, as she made no reply. "I'm afraid I will be unable to speak freely to Josiah, because my replies to him will become misaimed attempts to explain myself to you."

He had unwittingly calculated rightly by placing his dilemma in the context of something that concerned Josiah's welfare.

"Very well, Professor Sullivan," she said, looking up. "At the first sign of foolishness, I will try to tell you what I think."

Martin knew he'd been checkmated. He had confessed desire to know what she was thinking, and she had parried with what seemed a concession. But he only saw her in Josiah's company. And in their strange household, Helen was not asked her thoughts, though he and Josiah discussed matters that must be central to her intellectual life. Or once were.

He walked her to the door, but just before she stepped over the threshold she stopped. She turned and held out her hand, as though they had completed a business transaction, were it not for her earnest gaze as she spoke.

"Thank you, professor."

The firm pressure of her fingers against his hand communicated much: this was what she had come so far to say. He didn't even mind the "professor." When she was gone, Martin went back to the parlor and sat down, unable to tell whether the effect of this conversation on his frame was exhaustion or exhilaration. Pressing his right hand against his knee, he thought how strange it was that the electricity in someone's fingers could linger even in their absence.

"Well? Who was that?"

His father had entered the room. A vague sense that he would lose the sensation pressed into his palm inspired more shortness to Martin's reply than usual.

"A friend."

That only made his father more provocative.

"Pretty friend, Marty."

"Is she?" Martin said absently.

Eventually he looked up and registered his father's searching gaze.

"Reverend Gray's daughter-in-law," Martin said. "She's been helping us with his collection."

"Well! I'm not sure who I should worry about more," Will teased. "The Wobbly's wife or the minister!"

Martin said nothing, so his father changed the subject.

"How's your head?"

"Better."

"Good. I'm off to the Seattle Labor College. Professor Kincaid's giving tonight's lecture. Want to come? Though your friends at the IWW won't approve at all."

He pulled a paper out of his pocket and handed it to Martin. It was a pamphlet describing a program of lectures for which Trevor would provide the "zoological" viewpoint. The title page read "Knowledge is Power." He knew Trevor spoke to community groups quite often, but hadn't known the Seattle Labor College was on his circuit. Frank had ridiculed this group for stifling revolution with visions of the inevitable evolutionary progress of society. He handed the pamphlet back and shook his head. He didn't tell his father this, but he was done with labor politics for now, "zoological viewpoint" or no.*

Or so he thought. The next morning a gentle knock on the side of the collection room door alerted Martin to Trevor's presence. Usually Trevor took the liberty of sitting down for a while when he stopped by. Now, he stood awkwardly in front of Martin's desk. Martin thought he must have a quick query and did not set down the pen with which he was composing a set of lectures on Comstock's method of creating an evolutionary classification.

"That's a nice wound you've got there."

Since Trevor had not asked where he got it, Martin said nothing.

"Rumor has it you were seen at a Wobbly meeting," Trevor added.

The question was abrupt, as though he had to force it out. Martin put his pen down, sat back in his chair, and looked at his boss, who he now realized looked quite uncomfortable.

"Who cares so much how I spend my evenings? I'm pretty confident it's not you."

Trevor relaxed a bit.

"No wonder you haven't kicked me out of your office," he sat down. "Here's the thing, Martin. We have enough trouble with rumors the entire faculty's communist. Suzzallo's support for the eight-hour workday during the war set the sights of some powerful men on the university. There's no reason to bait them with tales of one of us touring the Wobbly lecture circuits. Especially if it's not true. Your dad's politics are pretty well known, of course. But one can't always determine the politics of a son by his father's. A man's just as likely to rebel as go along."*

"Dad said you gave this week's lecture at the Seattle Labor College," Martin said. "Surely you're on somebody's list too."

"They're considered pretty tame by comparison." Trevor shrugged. "And Ben's been after me for years to do my civic duty."

He added a wink. They both knew Ben despised the labor unions for ignoring, as he put it, man's selfish, competitive instincts and imagining that regulation and equal pay would create an earthly utopia. Martin smiled but the joke died away, and then Trevor hesitated.

"I better lay it all out. Yellow's tolerated around here. Slow reform and all that. But not Red. And rumor has it a few students are attending meetings."

Martin stared at Trevor's necktie. In the hours trying to recover his memories of the meeting, he had placed the familiar young face: a quiet anthropology student who had attended the previous summer's session at the biological station. Trevor took a deep breath.

"I'm supposed to ask you for any names."

Martin looked him in the eyes but said nothing.

After a moment's silence, Trevor added, "I suppose if you could just tell me whether you'll be attending any more meetings in the future."

"I don't mind telling you that I've no other meeting on my calendar," Martin replied. "But I'd rather that fact remain between us. If whoever cares is willing to ask me directly, then I may or may not decide to tell him the same."

Trevor nodded. He stood up.

"I'd no idea we're watched so closely," Martin confessed.

"Well, usually we aren't. Not as much as the humanists anyway."

Two days later, without any hint of blame or appeal that he change his mind about not turning informant, Trevor told Martin that Weyerhaeuser had decided not to fund the museum.

Martin shook his head. "I'm sorry, Trevor. I guess I called their bluff wrong."

"A man mustn't apologize for his convictions, or the power of his silence, in a case like this," Trevor replied firmly, then paused before adding, "We need to tell Reverend Gray. Would you like me to call him?"

"No. It's my fault," Martin answered. "I'll do it."

"I wager they were casting about for an excuse to spend the money on a genetics lab instead. You were just kind enough to give them one sooner rather than later. I don't think they ever understood why I was asking for a museum building in the first place. Not easy to explain to businessmen. Never has been."

He fell silent a moment, and then, as though, remembering something, his tone changed.

"You know, we need a minister at the station for this summer's session. Reverend Nye can only come for a week and I want someone for the entire session this year. Do you think Reverend Gray would come out of retirement? If he stayed in the room off the dining hall he could learn the lay of the land more easily. Might be a good holiday in apology for our about-face. All expenses paid, of course."

After Martin had agreed to convey the invitation, Trevor paused at the door and turned back with a vague smile.

"And we were worried about the outcome of Bryan's campaign."

Martin sat at his office desk for some time without moving. Trevor had spoken of his convictions, although he was infamous on campus for refusing to speak on issues other faculty debated for hours. Trevor had implied Martin took a stand by remaining silent in the face of those who would police a man's movements. The practical outcome of that action was clear: he had lost the collection.*

Apart from a violent slam of his desk drawer when he forced himself to move and get back to work, no one would have guessed the loss had they entered his office a few moments later and observed him bent over his notepad and Comstock's textbook once more.

At the next department meeting, Trevor casually mentioned that Weyerhaeuser had changed his mind and pulled the funds for the museum. Ben, who had sat visibly bored amid various bureaucratic tasks on the agenda, looked up with an extraordinarily sympathetic oath and a rapid-fire interrogation.

"What the hell? Sullivan's spent almost two years working under the assumption the university would keep that collection! He's wasted a year making trips to Tacoma to inventory that old minister's collection!"

"It wasn't a waste of time," Martin spoke up, but Ben was having too much fun being indignant on Martin's behalf, and it was a while before he let the subject drop. Despite all his warnings that Trevor and Martin best pay more attention to politics and the man on the street, Ben didn't add a single "I told you so." Instead, he bewailed the fact industrialists and businessmen thought they could tell scientists what they should be doing with their time.

By the next day, Martin's resigned return to his desk had driven Pete to distraction.

"This is maddening!" he exclaimed, throwing himself down in the chair near Martin's desk.

Martin looked up from his microscope. "We're going to try and get you a salary line from somewhere else, Pete. Trevor has some ideas."

Pete made a dismissive movement with his hand. "I don't care about that. I mean, I do. I appreciate him trying. But . . . the collection! Doesn't this infuriate you?"

Martin shrugged. To such a query from Ben, he would have followed the movement with a disciplined silence. But he did not mind discussing such things with Pete. He had been thinking a lot lately about what determined an individual's ability to do certain kinds of work.

"Sure it does," he replied. "I just don't expect another man to see things the same way as I do. There are plenty of entomologists who think that getting rid of the old museums will be a good cleanse of our obsession with dead specimens. That it will force us to pay more attention to living animals. That ecology will finally get its day. If you've been waiting for such a shift, then Mr. Weyerhaeuser made the right decision. Applied entomologists have been insisting for years that those who study life history are more valuable, more ethical even, than taxonomists, since their work helps horticulturalists and farmers. From that point of view, our patrons are simply catching up to what is good."

"But who's going to pay for accurate classifications?" Pete pressed, trying to raise the stakes to something Martin cared deeply about. "The applied men already argue you don't need to know all the species to kill the annoying ones. And goodbye to the foundation required for studying evolution!"

Martin picked up a tray of specimens and stood to place it back in one of the cabinets.

"Maybe it's been a fool's crusade to think we can amass enough specimens to study evolution in museums since the war. The geneticists collecting live drosophila in milk bottles and feeding them on cheap bananas to study variation have a great advantage over us."

Pete watched him in silence as he returned to his desk. "You forgot to say, 'If the work is done well.'"

Martin sat down at his desk again with a resigned smile. "Yes, Pete." he said. "If the work is done well."

Part IV

The Historians

Chapter 13

In the opening months of 1925, public debates between defenders of science and opponents of evolution were broadcast by radio up and down the West Coast. Ordinarily Martin wouldn't have known this outside of reports from Ben, but during his next trip home, Will purchased a radio and installed it on the table that already served as desk and dining area.

Martin hated that bulky wedge of modern life, though in the interest of domestic harmony he said nothing. It took up a third of the table space and all of the silence that had existed in the pauses between Will's newspaper clippings. It meant he had to listen to the Baptist minister from Minneapolis William Riley warn that evolution would result in a godless world of adultery and theft, since "in the struggle for existence" only the stronger have rights. He had to listen to the Seventh Day Adventist political scientist Alonzo Baker declare that according to Darwin's disciples, Christ could not have died for the sins of men, for in their view sin was nothing but "the hang-over from our animal ancestry, the remnants of the tiger and ape in us." The Baptist minister from New York John Straton joined the radio waves to denounce belief in evolution as an atheistic religion. And if the teaching of the Bible and Christian religion is to be ruled out of public schools to uphold the separation of church and state, Straton thundered, then anti-Bible and anti-religious teachings must be barred as well.*

Then Martin had to listen to the science writer Maynard Shipley try to answer them. Shipley recounted tragic tales of the Church's persecution of Galileo and Columbus, and argued that public school curricula must be left to those best qualified to judge by education and experience: trained scientists. Shipley argued that evolution should be taught because it supplied students with sanctions for right conduct based, not on someone's idea of what constitutes right and wrong, or what someone tells a man is good or evil, but on the immutable, unavoidable laws of nature. "From the unchanging operation of these laws no one can hide," he warned. "From the consequences of violation of these laws none can escape."*

Reports on developments back east were also constant, thanks to Ben. He reported that Reverend Straton had posted a warning on the door of his Calvary Baptist Church in New York City: "Is the American Museum of Natural History Misspending Taxpayers' Money and Poisoning the Minds of the School Children with False and Bestial Theories of Evolution?" Henry Fairfield Osborn, director of the museum, was firing back in lectures, books, and the pages of the *New York Times* and arguing that every act of creation, whether via evolution or no, is an act of God. The entire contents of the museum, Osborn insisted, demonstrated that creation happened via evolution. Ben didn't like the strategy, but included several versions of "At least he's engaging!" in his attempt to rouse his colleagues during department meetings.*

Normally a trip to Tacoma to work on Josiah's collection would have been a welcome escape from such drama, but now Martin had to somehow tell Josiah that, as Ben put it, all their work had been in vain and he'd put up with Martin's weekly visits for naught. Martin knew Josiah wouldn't see things that way. They had each gained a friend. But it was still heartbreaking news to report, after all those hours inventorying Josiah's beloved specimens in preparation for a new home where they might be studied in future.

He made the trip on a cold day in January, arriving before Josiah came down from his afternoon nap. He wanted to ask Helen about Trevor's invitation to Josiah to serve as chaplain for the summer session at the biological station first, and had resolved to promptly rescind the invitation if she betrayed the slightest hesitation.

She opened the door with her usual "Hello, professor" and took his coat and hat. Martin had hoped she might at least abandon the "professor" now. As she turned to place his things on the hook by the door, he tried to reconcile himself to their old pattern. He had a vague sense that otherwise this home would become something of a torment.

"I came a bit early because I need to ask you something," he said. "About this summer."

He spoke the latter quickly, so she need not fear he was going to demand she share her thoughts, about Nature or God, so soon. She followed him into the study and stood beside Josiah's chair, waiting as he set down his satchel. So he delivered Trevor's invitation, adding as he finished, "It will demand many of your moments. He'll be in an unfamiliar place. So I wanted to ask you first."

"I'm sure he'll be delighted to go. It's very kind of you, and Professor Kincaid."

Martin hadn't meant to tell her about the loss of the museum first. But when she took the invitation as an act of kindness, he said, "I'm afraid we're offering the trip in consolation for bad news. The funding for the museum's been pulled. We can't buy the collection after all."

She sat down in Josiah's chair with a look of concern in her eyes, which meant Martin was confronted with the memory of the last time he had been in this room, when she had gently pressed her hand to the side of his face and asked whether he knew who she was. This was his first hint that the warmth he had felt in Mrs. Macleod's sitting room had not been the result of the radiator in the corner.

"I'm very sorry," she said. "What happened?"

He didn't realize how long it took him to reply. Nor how she interpreted his pause solely as distress at the loss of the museum.

"Priorities change," he said finally. "I'd been warned to take more account of the business side of things."

He almost added "political side" but caught himself.

"Mr. Weyerhaeuser changed his mind?"

Martin nodded. "We're supposed to keep any insects that are important to the timber industry." He smiled slightly. "As though we can easily tell."

"But what are you going to do without a collection?"

"I haven't figured that out yet. Melander might be willing to buy the ichneumons, since some are good for pest control. If he does, I could take a closer look at them."

She smiled. "Oh, I wouldn't do that, professor. They might finally convince you of something."

Martin's ability to reply faltered under this reminder that she had been listening so closely, and for the smile that accompanied the "professor." As though it was a joke between them that he had the title at all, when he could pronounce on so little. She stood up and went to the cabinet to retrieve the trays they had been working on before Frank's visit interrupted everything. He knew that once she placed them on the little table in front of him she would leave. So he permitted himself one question. She, after all, had asked him three.

"Had you read *The Descent of Man*? Before reading it to Josiah?"

She replied with a simple no and he feared she might be annoyed at his blatant breach of the patterns of the house, in which she was not asked questions. Then she looked at him, and with a playful turn to the corner of her mouth, she spoke. "But I've read plenty of medieval versions."

The crooked smile did something to her eyes he hadn't witnessed before. She had a sense of humor.

He took a deep breath and said, "I was right. There were barbs coming from your corner." He crossed his arms over his chest and leaned forward. "All right, then. Let's have it. Only fair."

So she sat down again in Josiah's chair and told him a few tales from a medieval book of beasts. Of the lion covering his tracks with his tail so that the hunter could not follow, and how that story taught that Christ initially hid his divine nature from unbelievers. Of how the ant-lion, born of the lion and the ant, had two natures and was unable to eat meat or seeds, and perished miserably just like every double-minded man who tried to follow God and the devil. Of how the fact that even the brave antelope, with its great, strong antlers, was sometimes ensnared in brush so that it died taught men to beware the snares of the devil and that the wise man flees women and wine.*

"That sounds like a strange kind of natural history," Martin said when she was done.

"Yes. It does. The observations were often wrong, and the lessons sound superstitious. But the stories they told about animals convey something. About their hopes and fears. They might even make sense, not as science, but as stories about nature that were coherent and useful. They taught something about God and what humans must do to be saved." She paused. "Sometimes I think we best not be too hard on them. A hundred years from now we might look quite medieval ourselves, when this book," she gestured at *The Descent*, still lying on the table between them, "has had its moment and the world moves on. I wonder what will be written of the men who talk so blithely of superior and inferior races. Or whether socialism or capitalism is more natural. Or of what the disposition of a bull might say about the intellectual capacities of men."

Martin stared at the set of specimen trays on the table.

"Those are damning accusations," he said finally.

"Only if one thinks scientists are better than other men. Otherwise, I'm only stating the obvious. That you are human beings, and sometimes see what you wish to see. If it makes you feel any better, I don't think historians are any better at the task. We're quite good at coming up with stories about the past that support whatever cause or campaign we believe true in the present."

"But I know men who despise biologists' use of nature to justify what they desire or believe," Martin said. "Who walk away if anyone starts using ants as an argument for socialism, or the differences between stallions and mares as an argument against female education. Some biologists are more careful."*

"I know, Martin."

Somehow, with those two words, and her use of his first name, the threat of barbs disappeared. They heard Josiah's footsteps and cane on the stairs, and Helen stood. Josiah had heard his voice and entered the room with an outstretched hand and a "Martin!" When Martin took his hand Josiah wrapped his other arm about his shoulder in an embrace.

"You've returned!"

"Yes, of course I have," Martin replied quickly.

"You aren't angry with us?"

Martin glanced at Helen, as she took Josiah's cane and set it by his chair, but she did not look up. Then she left the room.

"No, Josiah. Why would I be angry?"

Josiah just gripped his arm and said nothing. Martin thought when Josiah settled in his chair that they would take up their work. Instead, his friend was silent, as though listening for something. After a moment he leaned forward and said in a hushed voice, "I thought he came to give her a divorce. But she tells me the only grounds for divorce in New York is adultery."

Despite the mysteries that pressed on his mind—Why were they married in New York? Why had they separated if not adultery?—Martin did not want to talk of Helen in her absence. He tried to shift the conversation to a topic on which at least one witness was present.

"I don't understand, Josiah. Your estrangement from Frank, I mean. You don't dismiss a man because he thinks differently than you. Even if he's an atheist."

"No. Our views of God did not separate us. Not directly. It was what should be done for his wife. For Helen."

They both heard Helen's footsteps as her name lingered in the air between them.

"Well, what tray are we on now?" Josiah said, leaning back in his chair.

When Martin told him about Weyerhaeuser's decision Josiah seemed more intent on alleviating Martin's feelings of regret than bemoaning the loss of a home for the collection. He asked that Martin still take charge of finding his lady beetles and wasps new owners. And he was thrilled with the invitation to the biological station, peppering Martin with questions about the students, and how Trevor thought a minister could best serve them.

"Just be on hand in case you are needed, I think," Martin replied. The memory of Connor's body on the rocks overwhelmed him for an instant before he forced the image away. "A short service on Sunday for the students who want one, perhaps. I'm actually hoping it will be a nice holiday for you both."

Later, after they sat around the dinner table, Martin noticed a feeling in his gut that he had never felt in this home before. He lost the bent of Josiah's conversation completely when he realized it was anger, as Josiah continued their old pattern of talking as though Helen had no interest in their conversations. He could not reconcile his friend's behavior toward his daughter-in-law with the liberalism with which Josiah faced so many ideas.

Sitting on the deck of a boat late that evening, he tried to subdue the anger by telling himself he had no right to question the patterns of his friends' household. Not if Helen accepted them. But over the course of the slow journey home, he decided that fact did not mean he must partake in those patterns. He knew very well that rebellion against Helen's routine of silence would require actions that might be read as courting. But he had resolved to apply his father's rule: that to hesitate from fear of being misinterpreted would exclude Helen because she was a woman. So he decided to risk a firm no before giving in to this household's habits again, and came an hour earlier than usual the next week as well, this time with his own, marked copy of *The Descent of Man* in his satchel once more.

She seemed content for him to work on the trays of ichneumons alone, as she carried out the various tasks for which Frank had ridiculed her. Having come into the study to retrieve something from her worktable, she stopped.

"What's this?"

"Something I wanted to ask you about," he answered. "It's underlined. At the bookmark."

He heard her step toward him. She sat in Josiah's chair with the book in hand, and opened it. Martin had not expected her to read the passage aloud.

Man is the rival of other men; he delights in competition, and this leads to ambition which passes too easily into selfishness. These latter qualities seem to be his natural and unfortunate birthright. It is generally admitted that with woman the powers of intuition, of rapid perception, and perhaps of imitation, are more strongly marked than in man; but some, at least, of these faculties are characteristic of the lower races, and therefore of a past and lower state of civilization.

He had struggled over which passage to place before her a second time, as a means of asking what she thought of the conversations to which she had listened in silence. He might have chosen a less damning one, but he wanted her to know he was not afraid to be told that one of biology's heroes had not looked very closely. But to hear this particular passage in her calm, quiet voice now, with Frank's admonishment that she could tell him a thing or two resounding through his head, was penance indeed.

"I've been trying to imagine what it must have been like for you to read this," he said when she was done. "And to know it has been taken as fact."

She looked up at him. "Little different, I imagine, than what it must be like for a man who is supposed to be of a lower race to read it. Or a man who does not, in fact, delight in competition."

Martin did not know what to say to that, so he said nothing. She looked down at the book again and turned the page.

"But what you or I feel isn't supposed to influence our assessment of its truth, right?" After a moment she closed the book and gently turned it over in her hands. "No bearing on whether he's right or wrong. Or whether he's got the facts right."

Martin's brow furrowed. Was she teasing him? Because this was the line he had taken so often with Josiah?

"All right then," he said finally. "But what about as an historian? As someone who studies the past, and must judge whether the stories we tell are true or false?"

She did not reply. So he held out his hand for the book.

"You must have decided whether this is good history, whether you like the conclusions or no. Listen. He's making a claim about the past. And the historical causes of the present."

He read the next passage aloud.

The chief distinction in the intellectual powers of the two sexes is shown by man's attaining to a higher eminence, in whatever he takes up, than can woman—whether requiring deep thought, reason, or imagination, or merely the use of the senses and hands. If two lists were made of the most eminent men and women in poetry, painting, sculpture, music, history, science, and philosophy, with half-a-dozen names under each subject, the two lists would not bear comparison.

When he stopped and looked up she was looking at him. Into him. And she did not answer his question.

"Why are you doing this?" she said with a tone that verged on sternness.

He might have said, "Doing what?" He might have asked what she meant. But he knew perfectly well. So he tried to look deep into his own heart before answering. He tried to match the sternness of her voice by speaking firmly. With certainty.

"Because you are a human being who has thought about these things. And because I was wrong to assume you had no interest in our discussions."

She released him from her gaze but was silent. Martin convinced himself in that moment that he would be refused. Then she spoke.

"He's assuming that the present state of things is due to nature rather than environment and education. He took the easy route. And no. That isn't good history. But perhaps attributing the differences between men and women to nurture would concede too much to those who would separate humanity from the rest of creation. Because it appears to place them outside the bounds of natural law and natural history. Perhaps he's just being single-minded, and has no intent to oppress half of humanity. That's your line, isn't it?"*

Martin said nothing, staring at the open book in his hands.

"Some thought a very different message might be wrested from Darwin's work," she continued. "Despite such passages. Some of the more radical suffragettes thought that Darwin had emancipated women because his history turned the fall of man and everything justified in its name into myths. That this book would complete what higher critics had begun, and remove all the biblical justifications of placing men and women in separate spheres."*

Martin finally found his voice, because he could ask a simple question.

"Higher critics?" he said. He knew the phrase only vaguely, for all his dad's tutelage about the triumph of scientists over theologians had justified ignoring it.

"They were theologians who began reading the Bible as an historical document rather than divinely inspired. They worked a generation before Darwin, trying to make sense of the inconsistences and understand the stories as products of particular cultures and societies, in the hopes of discerning the Bible's true meaning. They had one guiding premise in studying the past: that natural laws and human nature have not changed. No exceptions. So. Jesus was a man. A good man. An extraordinary man. But a man nonetheless. Because the chains of causation, God's laws, must never be broken. In a way, the higher critics developed the methods of modern history. Both yours and mine."

"And paved the way for Darwin?"

She smiled. "Maybe. There must be some connection."

"So they're the ones who declared the miracles, the virgin birth, the atonement, the resurrection to be fables?" Martin said, trying to remember Reverend Harrison's list.

She nodded. "Destroyed them all. The orthodox versions anyway. The literal versions. But remember they were Christian theologians. Every single one. Theologians who couldn't see how a religion that continued to rely on miracles could survive in a scientific age. They feared Christianity's great moral insights and power to do good were doomed, if men and women continued to insist on reading the Bible's miracles as literal accounts of the past."

"Then you agree with Josiah?" Martin pressed. "Bryan's picked the wrong target when he attacks Darwin? And biologists?"

She smiled again. "No. I'm quite sure you should be in his sights too. Attempts to modernize Christianity may be Bryan's ultimate target, but he knows he has to go through Darwin to meet them. Darwin's the one who gave those who wanted to believe God governs through natural law, rather than intervention, a natural explanation of these." She gestured to the specimen trays. "David Strauss was one of the most important higher critics. He loved Darwin's work because he thought a natural explanation of apparent design opened the door by which humanity could finally dispense with the superstitious tales that kept men bound in orthodox tradition. That men and women might see themselves as beings that have risen higher, by their own efforts, rather than as the descendants of a couple created in the image of God who were kicked out of paradise for their depravity. Strauss's history of humanity would remove the need for clerics. A church of state. And any intermediary between individuals and God." She smiled slightly, "That's political tinder, you know. At least, it was in the 1840s."*

He knew he should let her get back to work. But he asked one more question, because he might not have another chance.

"I thought you studied the Middle Ages. Not recent history."

"I did. But studying a time in which Christianity was so different makes you wonder why it has changed so much. The medieval churchmen didn't believe insects like this were created by God." She gestured to the ichneumons again. "Flies, gnats, and mosquitoes were the result of the corruption of nature after the fall of man. Of sin entering the world. Even the ichneumon wasps would have made a great deal of sense to them. The Bible says, 'The whole creation groaneth and travaileth in pain together until now.' Perhaps they had more cause to remember those words over the ones that imply nature demonstrates God's goodness." She paused. "I don't think they'd have had much trouble acknowledging the struggle for existence. There were orthodox Christians in the 1860s who thought Darwin's view of nature a fine description of the suffering and pain that entered creation at the fall of man. But Josiah is right. Somehow men and women's expectations of nature and of God changed."*

So she had been in the room still. When he had read the passage from Winwood Reade aloud, that insisted struggle and death as the means of progress could not be the creation of a good God, that therefore God must not exist, and prayer was hopeless.

She paused, before adding with a smile, "That and I used to practice my German by reading Strauss. He's easier than Hegel."

Martin smiled too. He had been subjected to Hegel for six years when he was learning German. He learned a great deal of vocabulary and could read scientific articles well enough, but trying to understand Hegel's philosophy had always driven him straight back to his beetles. Before he could reply, they heard Josiah's voice come from the top of the staircase.

"Helen dear, do you smell something burning?"

"Oh! The pie!"

She rushed to the kitchen and Josiah came down the stairs.

"That isn't like her," Josiah mumbled to himself as he entered the study.

Martin realized he had interfered with her preparations for dinner. He knew from his father's constant talk of the burden placed on housewives how much work she had to do during the day. Neither Helen nor Josiah had mentioned a housemaid, though a home up to this standard usually had one or two. Which meant Helen did everything.

During his second visit months ago, he had insisted on taking over the task of stoking the fireplace when the embers burned low. Now, he wanted to ask what he could help with so she might be able to talk longer, without worrying about a pie in the oven or housework. But he couldn't figure out how to word the request. Then he remembered Frank's claim that she didn't like to cook. So the next week, after an hour at the Experiment Station, he picked up several cans of strawberries from one of the few farm stands that remained open in the winter in Puyallup. Rather than handing them to her at the door, he offered to make a pie for dessert if she didn't mind letting him use her kitchen.

He worked alone before he had to interrupt her own tasks to request some hints as to where she kept the flour. She found him a container and placed it on the counter. "The sugar and cornstarch's also there," she said. "Do you need a lemon?"

"Brought one," he said, gesturing to the satchel he'd placed on a chair.

She disappeared behind a door near the icebox and from the echo of her footsteps Martin could tell there were stairs going down. After a moment she returned with a bowl of dried beans. She sat down at the kitchen table to sort out the pebbles and asked whether his mother had taught him to cook.

"No. Dad's got pretty liberal ideas about what a man should be able to do. I was the only one he could apply them to."

She smiled as he glanced at her. They worked in silence as the berries simmered in their bath on the stove and Martin rolled out the dough. It wasn't until he had placed the shell in the oven, removed the berries from the stovetop, and stood at the sink washing out the mixing bowl that she spoke again.

"May I ask you something, professor?"

"Yes, of course," he replied.

There was a long pause before she said, "Do you ever fear that your tolerance of Josiah's views is disingenuous, when you don't believe we can know so much?"

Martin's hands became still for a moment. Then he drew in a quick breath, dried the bowl before placing it on the counter, and sat down in the chair across from her.

"I suspected you might have some hard things to say about biologists. I didn't expect you'd have such a hard thing to say about me."

"I don't mean it to be hard. But I promised to tell you what I think. And it's something I've often felt when speaking with him."

Something in her confession of unbelief, and that her silence did not arise solely from the rhythm of Josiah's days or what visitors expected of her, touched Martin deeply.

"You don't have to answer," she said when he still gave no reply.

Normally Martin would have accepted the offer of not replying with relief. But his mothers' lessons in paying careful attention to the beings around him, in the interest of justice, had become all mixed up with his knowledge that this woman, a scholar, had sat near for months in silence. And so, gazing at the yellow and white tablecloth, he shared something he had confessed to few friends.

"No, it's all right. I know some of my friends see my agnosticism as a lack of conviction. Or cowardice. But I can't make myself care enough to decide one way or the other about all these things that men and women believe or deny with so much faith. Perhaps I've repressed it too far, or it simply isn't part of my makeup. It's not even something I've ever really thought through. I can't give reasons for it. I mean, I could try, but they always seem rationalizations of beliefs formed God knows how." He paused. "But I do know that my unbelief is no reason to try to pull anyone down on either side. If I take agnosticism seriously, I'd not know on which side to fall."

When he looked up she was smiling.

"You remind me of the old theologians who insisted a man cannot know the mind of God. They believed it was heretical to claim knowledge of nature's laws because that would presume knowledge of God's mind and ways. And that the best fallen man can hope for is to describe a little bit of the nature at his feet. If he was humble and remembered his tendency to sin."*

Martin looked down at the floor. Ben had said he was like a monk, which was not, in Ben's world, a compliment. But Helen's accusation was sympathetic.

"You make it sound quite virtuous," he said after a moment. "Pious even. But caution isn't always considered a virtue by my friends. A ready excuse for being hidebound and old fashioned, maybe, since it justifies hours and hours of descriptive work. My work is generally seen as rather useless. Species making or stamp collecting. Not science."*

"Surely that's better than writing a bunch of nonsense about nature's laws?"

Martin nodded. "I think so. Problem is it doesn't inspire much respect or many patrons."

"But you don't care."

"No. I don't. I probably should. Well. Clearly I should. But whenever I try to make an argument that any of the work I do is useful, I remember that I became a naturalist because I fell in love with insects as a child. I can call it contributing to knowledge and all, but I'm not sure why a millionaire should give me a part of his fortune to do as I like. I don't mean to belittle those who do ask. I know good biologists who believe that the study of natural history is valuable because it creates a habit of mind that demands as exact knowledge as possible touching any situation in life where decision and action are necessary. And that sounds quite useful. I just don't understand how they know when to stop. When one can finally say I have separated the content and wishes of my own mind from what is true, and now act based on a knowledge of reality rather than illusions. I don't understand how anyone knows when to say, 'Here, we know enough. Now do this.'"*

He stopped speaking, a bit astonished he had spoken so much.

"I misspoke," she said. "You do believe in something."

Martin was unsettled by her candidness in speaking of what he did or did not believe, when he could barely determine his own convictions. But he was glad she didn't seem to mind his presence in her kitchen. At least, that is how he classified what he was feeling when he heard Josiah's footsteps and cane on the stairs. He took the baked pie shell out of the oven and poured the berry mixture in, before picking up his satchel and joining Josiah in the study to work through some of the last ichneumon trays.

Josiah had pointed him to some doubtful species designations, so Martin was spending a lot of time at the microscope, trying to determine whether the genital armature varied geographically. At dinner Josiah paused their discussion of Martin's latest findings after taking a bite of the pie. He turned his head in Helen's direction.

"This a new recipe?"

"Yes. Martin made it."

"It's delicious," Josiah said. "You should recruit him next week too. Not that you can't make a good pie."

Helen smiled but said nothing.

"What else can you do?" Josiah asked.

Martin laughed. "You make it sound like a party trick. What else would you like to eat? I'm good with potatoes."

Josiah laughed as well. "All right, then. Let's see what you can do with potatoes."

So during his next visit Martin brought ingredients to make his father's favorite potato dish. Then, as the weather turned and new fruits and vegetables began appearing at the farm stands, he brought something to contribute to dinner each week. At first, in the hour before Josiah came downstairs, and he helped Helen prepare dinner, they worked mostly in silence. The quiet moments meant he had time to notice how much care she took with each task as she cooked. It was how she cleaned too: methodical and focused. None of it matched Frank's words that she didn't like this kind of work.

He was surprised by how comfortable those moments of silence felt, perhaps because that was how they had dealt with each other for so long. But eventually he braved attempts at conversation, and even some questions. He learned about her dissertation on medieval bestiaries. And told her he had read the line, in the book by Eileen Power that Frank had given her, that despite the greater popularity and status of the sciences, no essential distinction existed between truth seeking in history and truth seeking in science. That both must proclaim certainties as certain, falsehoods as false, and uncertainties as dubious. The passage had reminded him, he said, of trying to say anything about the evolution of weevils with certainty.

Helen nodded. "Only our specimens are church registers, account books, inventories, and wills. We never have enough of them either, and have trouble seeing the ones we have rightly."

"Because scholars find what they wish to see?"

She nodded. "Or can only see certain things, whether they wish to or not. Most historians write about nunneries and monasteries as scandal-ridden prisons. But that's because we only have the records of the women who broke their vows of chastity. The good ones, the happy ones, the indifferent ones. They must have lived too." She gave a slight smile. "Maybe it's also hard for moderns to believe that women would choose such a life of their own free will. Or that they were virtuous enough to have lived up to the ideal."

She stood up from where she had been sitting at the table, across from him, snapping the tips off green beans.

"People have a hard time seeing that anything good may have existed within the walls of abbeys and monasteries," she said as she dumped the beans in a pot of water on the stove. "Or arose from them. Today men use them as examples of the deteriorating influence of religion on Europe. That's what Darwin thought, isn't it? That the men of gentle, thoughtful nature, given to meditation or culture of the mind, had no refuge except in the bosom of a church that demanded celibacy? And that the standard of intelligence plummeted as a result?"*

Martin was sitting at the table finishing an apple salad, but he paused and looked up. He knew by now that most eugenics textbooks contained sections lamenting modern women's tendency to delay marriage to train as nurses or teachers. As often as not, as evidence of the tragedy of such decisions, they told stories about the past. The darkest element of the "dark ages," claimed one, from a eugenic standpoint was the enforced celibacy of the priesthood, since this withdrew the best blood of the times from the population. It was, authors urged, an uneugenic custom that persisted in the popular assumption that scholars must live monkish lives of seclusion and contemplation. One text even argued that if this misguided ethic continued, then employment and attendance at colleges and universities should only be open to the eugenically unfit. Martin took shelter in the fact she was talking about Darwin, rather than his contemporaries.*

"That was Francis Galton," Martin replied. "He was citing Galton."

She laughed. "I'm afraid I must side with Josiah, professor. The point is Darwin's."

He had never heard her laugh before. He laughed too, though it was at his own expense. For the first time, Helen saw his smile spread to his eyes. She turned away and put the pot of beans on the stove. Some moments passed before she spoke again, and when she did it was to ask him to set the table, which sent him out of the room.

Given students were asking for extra office hours prior to exam week, it was three weeks before Martin could visit Tacoma again. But when he did, he brought a bunch of vegetables for a casserole. As Helen cleaned them at the sink and he waited to help with the slicing, she asked him questions about his students.

"What happened to the young man whose father wanted him to leave the university?"

"Pete? We asked the trustees to fund a position in applied entomology for him. Seems generous but it's a blow. He loves doing taxonomy for its own sake. But we can't convince the trustees there's a reason to map the boundaries between beetles that don't threaten any crops. About as difficult as it must be to convince them we should try and understand a time when men believed the earth was flat."

"That's not true," she said as she placed the bowl of vegetables on the table. "That scholars in the Middle Ages believed the earth was flat."

"No kidding?"

His father loved the tale of Christopher Columbus confronting a bunch of superstitious Jesuits denouncing the belief that the world was round. The story represented everything Will believed about the need for a man to buck authority and tradition, find out the truth for himself, and test his knowledge by action.

"It's often told, I know," Helen continued. "Ingersoll and White always insisted the medieval belief in a flat earth proved that ignorance was king because the Church governed so much, and tested all knowledge by appeal only to scripture. But it isn't true. Men in the Middle Ages knew the earth was round." She paused. "Maybe the idea medieval philosophers actually knew the truth about something doesn't fit with our belief in constant progress, or that we are so much better than our predecessors."*

"My dad raised me on White's *A History of the Warfare of Science with Religion*. And Ingersoll," Martin said. "They were like bibles in our house."

"Theology. Not religion," she replied. "He used the word theology."

Martin paused his work to look at her. True, it had been two decades since he'd picked up the book, but how could he get the title wrong?

"Did he?"

"*A History of the Warfare of Science with Theology in Christendom*," she said. "White didn't think science was at warfare with religion. He was just as troubled by the decline of religion as Bryan, in his own way. But he blamed theological orthodoxy and authoritarianism for the decline, rather than evolution. He, too, was trying to preserve Christianity as something a rational man could believe, so that its ethical truths wouldn't be swept away with the myths and legends. Why would a bunch of liberal theologians show up as heroes, if the book is about a warfare between science and religion?"

Martin thought of when she had caught him looking at the book on her desk, and he had told her of Comstock's constant "Look again. Look again." He smiled.

"I think my dad must have skipped those parts," he said. "Or have I forgotten them?"

"You're both in good company. Atheists ridiculed Professor White for his belief that the hand of providence is in the progress of history, while ministers accused him of infidelity for believing God governs by natural law. Both sides knew men and women might read the book and conclude that the Bible is a fable. But White thought science was purifying Christianity of its superstitious elements. That one could still find eternal truths in the Sermon on the Mount and the Lord's Prayer, even if Christ's miracles were myths. Some demanded to know why anyone should stop there. Once the task of tossing things out has begun. He lost friendships over that book."*

Martin's brow furrowed, and he suddenly asked, "How do you know all this?"

"I studied with Professor Burr," she replied without looking up from her task.

Martin forgot the task beneath his own fingers, sat back in the chair, and stared at her. George Lincoln Burr was a beloved history professor at Cornell. He was a good friend of his own mentors, Anna and John Comstock. Martin knew it was Burr who had collected the books from which White composed his stories of science triumphing over superstition and orthodoxy.

"When?"

"1912 to 1916."

Martin wanted to demand why she hadn't told him this before. For she had known since his arrival that they must have overlapped at least a year or two.

"Professor Burr attended the dinners at the Comstock's," he said instead. "I always liked him."

She nodded. "You're very like him, in some ways. He always needed to read one more book before drawing any conclusions. He insisted intolerance must be vanquished by enlightenment, but struggled with how to impose that belief on other men. White never had those kinds of doubts. They had a devil of a time writing a history together. White thought a man should write history to change the world by convincing men that science and reason is the only path to progress. Burr only wanted to understand the past. For its own sake." She paused. "You are similar in that, too."*

Martin sat very still. He knew she had been listening. She had said as much. But it was still a bit overwhelming to realize how much she had understood. About him. The fact that she knew so much about his troubles and ways, when he knew so little about her, suddenly felt very unfair.

"And what do you think?"

The words were out of his mouth before he remembered Frank had said them, when he accosted her at dinner. But this time she answered.

"I think studying the past should teach us humility. I think history shows us how intelligent men and women thought they were right about things we deem wrong. And that it's very unlikely that some of our most cherished ideas and crusades will be spared a similar fate."

Martin tried to ignore the quickened pace of his heart as she looked him in the eyes and spoke so earnestly. Then she looked down and began slicing vegetables again with an effortless quickness. He'd never seen carrots cut so thin.

"Your dad would probably say my view is a pretty convenient excuse to do nothing," she added.

"Yes," Martin smiled. "He probably would."

He got a good dose of his own medicine when she replied with a slight shrug and a calm "Maybe he's right." She reached for the sliced vegetables in front of him, added the carrots, and then stood to place the vegetables in the pot of boiling water on the stove.

He had not noticed that Josiah had stepped into the room. Which was strange, because the thump of Josiah's cane was quite loud.

Chapter 14

Martin had never been so aware, despite Sarah, Phoebe, and all his dad's training, of the barriers to some individuals' pursuit of what they loved, as he was that spring. He could guess at the forces that had led Helen to abandon her scholarship. He knew the president of Harvard refused to let the Zoological Laboratory hire graduates of Radcliffe on the grounds women should, by nature, get married and perfect the home. He knew some biologists insisted women's suffrage was interfering with the progress of the race. A woman's place was in the home, they said, where her biological virtues of caring for her children and husband would uphold civilization. And he knew that, in the view of many of his colleagues, the presence or absence of children must be the primary basis of any judgment of a woman's life.*

He knew all this because for weeks he had spent hours reading books on eugenics to determine whether he could teach a course on the subject. He read stacks of reports coming out of Charles Davenport's Eugenics Record Office, trying to get ahold of concrete data behind eugenists' claims that some women should have children and others should not. But something of his hatred of speaking of those not present to defend themselves made him recoil at the authors' confident judgments of individuals' moral and physical worth. The ERO's family studies were full of genealogical pedigrees that claimed to trace feeblemindedness, alcoholism, sexual immorality, insanity, and a range of other "undesirable elements" deep into the past of certain families. The conclusion to be drawn from these pedigrees was clear: all of these traits were hereditary. One study of a poverty-stricken region in Pennsylvania blithely concluded that a pair of feebleminded parents who had produced all normal children must have been ignorant of the real father. That author thought nothing of describing her poverty-stricken subjects as silly, defective, inferior, boorish, and ignorant, all in a scientific treatise urging that such "degenerate" men and women be prevented from breeding.*

Martin was becoming certain of one thing. He could not assign Davenport's family studies to a class without constantly questioning their assumptions and methods. And he was also pretty sure that was not the kind of class James Slater had in mind.

But it was Ben lecturing him on the work of the American psychologist G. Stanley Hall that finally made up his mind to tell Slater no. They were driving up to Samish Bay on a clear day in May to help Trevor with his oyster experiments. Usually Martin could handle Ben well enough when they were alone. With no students to convert, Ben wasn't as aggressive and tolerated Martin's pauses and silences better. But it was a long drive, which left plenty of time for Ben to cross Martin's boundaries with his gossip and teasing. The fact Ben had his sights trained on one of Will's good friends, economics professor Theresa McMahon, didn't help. He'd read a newspaper report that she had testified in Olympia in defense of equal pay for women.

"From a purely biological standpoint," Ben said as Martin drove them through the fields of Skagit Valley, "tempting women with a good salary from the duties of motherhood constitutes the worst antisocial, immoral result of these feminists' misguided crusades. It's the pinnacle of selfishness from the standpoint of biological ethics."

Martin frowned. "Pretty easy theory to propose with no women here to question it," he said.

Ben laughed. "Wish a certain lady was here to let me have it?"

"What are you talking about?"

"Miss Phoebe Bartlett." Ben smiled. "Just thought I'd give you another reason to miss her."

"There's nothing going on between Miss Bartlett and I," Martin replied.

"Too bad. Case in point though, isn't she? Fits a certain feminine type described by Professor Hall perfectly."

"Afraid I'm not up to date on the latest psychology."

He regretted the sarcastic confession in an instance, for Ben ignored the sarcasm and took the confession as an excuse for a lecture.

"Hall argues that at a certain age women who have not fulfilled their role as mothers follow one of three paths. They become either dollish and selfish, ambitious and intellectual, or morbidly selfless. I'm pretty certain Miss Bartlett's headed straight for path number 2. Or she will soon if someone doesn't get her to an altar."

"Yes, she'd rightly let you have it for that."

"And maybe she should," Ben said. "Still. I wager it's more likely that all these feminist economists are trying to remake industry on the false assumption that the human race isn't subject to the laws that regulate nature. It's all human pride, no different from the likes of Bryan and Billy Sunday. These utopias are going to cause a whole lot more suffering than happiness. They go against everything biology tells us. Every biological imperative."

"And since when does biology deal with imperatives?"

"Since the first living organism was compelled to seek the light!" Ben replied. "C'mon, Martin. You don't believe in the alternative."

"That doesn't mean I think we've got human nature figured out."

"We know enough about certain things. And in this case the stakes are too high for all your depends and maybes. Ambition for equality is condemning women who would be ideal mothers to leave nothing to posterity. But I'll tell you the real biological tragedy: the third type. The ones who channel their motherly instincts into all kinds of charitable campaigns and self-sacrificing martyrdoms, from nursing to orphanages. It's the women who transfer the devotion and self-sacrifice Nature intended for husbands and children to strangers and the welfare of mankind, who impoverish the heredity of the world the most."*

"You do realize the grounds on which you criticize women's decisions makes whatever you've supposed about their physiology more important than their hopes and dreams?"

Ben pounced. "And that sounds a great deal like there's a realm biology can't enter. Pretty useful stance when biology says something we don't particularly like."

Martin took a deep breath. He wished he could repeat Reverend Harrison's insistence that every soul is of value and precious for its own sake in God's sight. But he didn't believe in the underlying assumption of that criticism.

So instead he asked, "And do you trust scientists to sit in judgment, Ben?"

"Up to now we've trusted men and women to choose their partners in complete ignorance of biological laws. Anything's got to be better than that, surely. Biologists or psychologists. I don't really care, as long as they're paying attention to the science." He paused before adding, "Where else'd you have us look?"

Martin said nothing. Ben turned and stared out the window for a moment.

"I may be wrong, Sullivan," he said finally. "I know that. But I'd rather be wrong while trying to think scientifically and ground my understanding in biology than tell kids they are the way they are because of original sin and their destiny might be hell."

"You don't think it's just a new, biological version of original sin if we're all instinct-bound machines, predetermined to do wrong?" Martin countered.

"Good grief," Ben said with a dramatic moan. "You sound like William Jennings Bryan."

He let the subject drop, though, and for the rest of the drive rambled on about university politics. President Suzzallo was in trouble with the new governor, a timber baron, for having supported the eight-hour workday during the war. The governor's newly appointed regents seemed intent on Suzzallo's removal. But Martin wasn't paying much attention to Ben's commentary on Suzzallo's fate. He couldn't get Ben's challenge that he sounded like Bryan out of his mind. It accused him of drawing an arbitrary barrier between humans and animals because something within him balked at applying the theory of evolution to his own kind.

It wasn't until he was knee-deep in the frigid waters of Puget Sound looking for evidence that the oysters were spawning that he realized what had upset him most about Ben's speeches. Ben's pronouncements amounted to a harsh biological judgment of Helen's moments. As harsh as Frank's, though the criticism came from so different a source.

Though he had delayed the decision for months, the conversation with Ben transformed into a firm resolution to travel to Tacoma earlier than usual on Friday and tell Slater no. He had made arrangements to meet Slater in his office, and found the president of the university, Edward Todd, waiting for him as well since he would have to approve the contract.

"I grant a great deal of rubbish has been written in the name of eugenics, sir, but there's an important core that it is our duty as educators to teach at this college," Slater was saying to Todd as Martin sat down.

"I don't mean to seem too conservative," the president said to Martin apologetically, and then turned back to Slater. "But it's just that the trustees may look quite askance at this topic, James. We're a Methodist institution, you know. We must be guided by certain values."

"Plenty of upstanding Christian leaders have supported this work," Slater replied with obvious frustration. "Eugenics doesn't constitute the abandonment of Christian values or ethics. It asks that we apply those values to future generations. Enlightened ministers know that if we refuse to embrace genetics as a means of doing everything in humanity's power to improve human life, we turn our back on providence. Is it conceivable that the God of love and mercy would desire misfits, physical or mental, to perpetuate their misfortunes on to their posterity? We must continue to minister to the feebleminded and wrong willed who are with us, but we must not allow them to contribute to the race stream. Eugenics shows us the way towards applying Christian compassion to the unborn."*

The president was listening closely, but gave no indication he was convinced. Slater switched tactics.

"Three hundred and fifty colleges and universities have courses on eugenics now, sir. We really can't afford to be so hidebound. We'll look like the Catholic Church, denying in the face of all evidence that the earth goes around the sun."*

Though he didn't seem convinced by Slater's theological arguments, President Todd was willing to concede to his expert opinion on what was required for a good biology curriculum. He left a note outlining a salary. Martin did not want to embarrass Slater in front of President Todd, and had remained silent since he had not yet been asked his views. But when Slater passed the note to him with the numbers, he did not look down so that it was clear salary was not at issue in his decision.

"I'm afraid I can't do the class. I've spent a whole term reading, but I don't think I can teach the kind of course you want."

Slater frowned and asked why not. So Martin tried to put his concerns into words.

"I just don't think we know enough about what constitutes fitness to make decisions about sterilization or mate choice based on scientific knowledge. Maybe one day we will. Then we will have to decide whether it's right to use it."

"I understand, Sullivan. I know how ignorant we are. But don't you think there's a danger if we're too cautious to apply what we do know? Proving that our powers of doing good are limited in certain directions is a pretty feeble excuse for neglecting to do the good that we can. Think I or any other man would have stepped on the battlefields of France if we insisted on knowing the outcome with absolute certainty?"*

Trevor hadn't mentioned Slater had served in the war. Usually that was included at first description of any man who had been in uniform. Martin tried to will away the increase of his heartbeat that would lead to trembling in his hands. In this moment, his old physiological response to debate was infuriating. Because chances were this man had been through so much worse.

"This nation lost thousands of its best and healthiest men in the war," Slater continued. "Europe lost nine million. Nine million of its fittest men culled in the trenches. While the weak were left home to marry at will. Those are facts, Sullivan. Tragic, biological facts."*

Martin had read this defense of the campaign for racial betterment in plenty of textbooks. And he'd imagined rebuttals often enough. He wanted to ask whether Slater was referring to the progress of the human race as a whole or just Anglo-Saxons. He wanted to ask what he thought of those who argued it was the pacifists who refused to go that would breed—or was it train?—men and women toward a better, more peaceful civilization. But the image of this tall, almost painfully thin naturalist crossing the Atlantic prepared to make the ultimate sacrifice, silenced his questions.

Instead he said, "I'm not sure how to determine my own worth, much less that of anyone else."

Slater frowned. "We know that it is better to be healthy rather than unhealthy. To be intelligent rather than feebleminded. And we know we are duty bound to give our children the best chance possible of a happy, useful existence. Charles Darwin himself pointed out that hardly anyone is so ignorant as to allow his worst animals to breed. Yet when it comes to our own species we throw caution to the wind and let irrationality and sentiment rush in."

Martin suspected Slater was still mentally arguing with his boss, while perhaps he was still mentally arguing with Ben. In any case, his reading of *The Descent of Man* had given him more ammunition than he usually possessed when colleagues pronounced on what biology said about human beings.

"Darwin also wrote that our instinct of sympathy compels us to give aid to the helpless. He didn't think we could check our sympathy, even at the urging of hard reason, without deterioration in the noblest parts of our nature."

But Slater, it turned out, knew *The Descent* as well as Martin. "He also thought that the weaker and inferior members of society did not marry so freely as the sound. We know now he was wrong. And he was writing almost fifty years before the war."*

He sat back in his chair and leveled his steely blue gaze on Martin. "This nation asked for the bodies of thousands during the war in the name of the public and national welfare. How does the removal of the power of procreation from a small number of mentally deficient or diseased, physically malformed individuals differ from that democratic principle? Would you demand individuals sacrifice their lives for the greater good, but not their reproductive capacity, though they inject their defective heredity into future generations of American children?"*

Martin could not answer questions like this, so he said nothing. Slater drew himself up again and made a resigned motion with his hand, as though sweeping all his arguments away as irrelevant to the moment at hand.

"All right, Sullivan. You're obviously not the man to teach the course, and I appreciate your making that clear. But I haven't got time to find anyone else. What about an entomology class? Although I doubt the enrollment will be high. Certainly not as high as a more practical set of lectures."

"If you really want a practical course how about one on economic entomology? I'd still feel a bit of a fraud given the angle, but at least the organisms are my usual object of study. And I suspect I'll be doing more of that kind of work in future."

Slater gave Martin the slip of paper with the same salary numbers, adding with a hint of irritation that the president would probably be relieved. The course would start the week after the end of the summer session at the biological station.

Martin practically ran to Josiah's home in an effort to use up the increased adrenaline in his system. When Helen opened the door, she took his hat and coat as usual and told him Josiah would be with him in a moment.

"I'm sorry I couldn't help with dinner," he said, a bit out of breath. "I had a meeting at the college that took longer than I expected."

Before she could reply Josiah's voice came from the kitchen.

"Is that Martin? Maybe he can help!"

Martin followed Helen into the kitchen to find an electric washing machine in the middle of the room.

"Hi Josiah," Martin said. "What's this?"

"Frank sent it," Josiah replied.

Martin looked at Helen, kneeling down to pick a bunch of parts up off the floor. She looked exasperated. And tired.

"Boy's inconsistent as all get out," Josiah added. "Plunking down a bunch of money for something we don't need."

"He thought it would be helpful, Josiah," Helen said. "History's moved by machines, and all that." She placed the parts on the table. "You two go ahead and get to work. There are only a few more ichneumon trays to record. I can figure this out."

Martin followed Josiah into the study with a strange tightness in the pit of his stomach.

As a good observer, he probably should have seen the warnings. He saw them in retrospect well enough. Later he would wonder whether he would have stopped his early arrivals if he had paid more attention: to how often he wanted to ask what she thought about something, to how it felt to inspire her smile or laugh, to the regret he suffered now at having argued with Professor Slater and arrived so late, to the sense of displeasure as he walked to the study and left her in the kitchen alone. And to something in his gut when she came to Frank's defense that he could not quite name, but that felt dreadful.

He could admit to himself that he liked her a great deal. He was charmed by her obvious sympathy with the loss of the collection, and by her attention to his own troubles and ways. He was proud of the fact she seemed at ease with him, though she still did not join his and Josiah's discussions. So he resolved the vague sense he was in danger sternly away, and fooled himself into thinking that, hours later, after they'd worked through the last tray of ichneumons and Josiah had gone to bed, he stayed a few moments more solely for the sake of repinning some dead lady beetles that had become dislodged from their proper place in their trays.

He had become so used to the silence in the room that he startled when she spoke.

"You're going to miss this work."

He looked up and saw her standing in the doorway.

"Yes. I am," he said. He looked down at Josiah's trays. "It's an extraordinary collection. I'm glad I got to spend time with it. There aren't many collections out here like this. Or back east. A generation ago we could gather thousands of specimens from all across the globe and revise conclusions that naturalists originally drew from too few specimens. But the world's changed."

"The war?" she asked.

"The war. Yes. Partly. Last year Harvard received word that the socialist colonial secretary of Jamaica refused to let collectors onto the island. Wanted to know why he should allow the fauna of his country to be carted off wholesale to gratify the passing whim of a moneyed class in another hemisphere. I don't blame him. The work's always been a bit strange, and it flourished in a certain kind of world. Perhaps that isn't the kind of world we want." He stopped speaking for a moment, afraid his commentary would be too transparently a response to Frank's criticisms. Then he shrugged. "But we had a good run since Linnaeus. Maybe it's best to give up the illusion we can get good work done, now that the old collecting networks are gone. And everything scientific's supposed to come from a lab desk."*

"It must have been hard to leave the museum in Cambridge," she said. "Surely you had access to more specimens there."

"Yes. But I didn't have much choice." He paused for a moment, and then confessed something he had never told anyone. "Dad got into some trouble and we had to get out of town. Sounds quite dramatic, doesn't it? He's always had a weakness for grand campaigns to change the world. During the war it was pacifism. We were constantly afraid he wouldn't be coming home, murdered by some patriotic mob. After the war he got on a bunch of lists as a communist. Then things died down and he assumed the witch hunts were over. But they weren't. Law firm finally had enough and fired him. He took refuge, or maybe revenge, in the arms of the wife of his boss." Martin took a deep breath before looking up and adding with a slight smile, "That, I suppose, is another one of his weaknesses."

She had come into the room as he spoke and was sitting in Josiah's chair now, her brow furrowed as she listened.

"But that isn't usually a weakness that gets a man chased out of town these days, is it?"

"Maybe not. But his enemies had some rather unscrupulous friends in the bootlegging business. Two days after I got the offer of a job here, a man showed up at our door with a pistol under his coat and a very polite warning that Dad best disappear. He doesn't know about that part. He'd have insisted on staying for a fight. And gotten himself killed."

"You're very different," she said as he fell silent. "From your father."

He nodded toward the specimens before him.

"This is my way of rebelling, I guess. My dad's very certain about many things. Maybe I love this work because it's always provisional. One needn't ever declare something is known for certain." He paused, staring at the tray of specimens before him. "I suppose that's not good enough anymore."

"But you believe in the work, don't you? It will be hard to leave it?"

Martin pressed his lips together in response to this gentle interrogation. For weeks he had let her ask him questions while only asking a few questions of his own. Frank's admonishment—that he studied with such care insects, creatures that had nothing to do with him, but refused to interrogate the human creatures nearest him—echoed through his brain. He nodded briefly in reply to her questions, looked up, and this time let the words escape.

"And you? Was it hard for you to leave your work?"

He saw her gaze falter and she looked at the floor. For a moment he thought he'd asked too much, but then she spoke.

"I was in London when we heard the Germans had burned thousands of medieval manuscripts in the library of Louvain. There were never very many to work from. Our work is similar for that reason, too. A few manuscripts to piece together the thoughts and hopes of generations over a thousand years. And the fewer the records, the more fanciful the tales men make. It was terrible to think we lost so much when so little remained. But it's strange, isn't it, to lament the loss of a bunch of medieval manuscripts during a war that cost millions of lives? Sometimes it's hard to believe that kind of scholarship is useful."

"But you said history teaches humility. Surely trying to get outside the content of our own minds and prejudices, using those remnants of the past, is of some good, even when the world is engulfed in hate. Perhaps especially then. Whether we're studying insects or medieval manuscripts."

She shook her head. "But that is precisely when we seem incapable of seeing things rightly. Think of how many explanations historians have offered for the war. The man who insists it was all the result of ideology and nationalism is countered by another who insists they were fighting for control over resources in Africa. The man who blames Nietzsche or Darwin is met with detailed arguments that their ideas had nothing to do with why men descended into the trenches by the millions to slaughter each other. When in the end the only thing we can know for certain is that it happened. And the world went mad."*

He had to say something lest her words hang in the air too long.

"We can go you one better and farther back in time. I have colleagues who insist war is an adaptive response to evolutionary pressures, and others who insist men are by nature peaceful and war's the result of an economic system that is out of step with our biology. Few seem to think biologists might have nothing useful to say about the matter." He paused. "It's strange. We imagine we can dispense with our hopes and dreams in understanding human origins, when even scholars of recent human history can't manage it."*

She gave a slight nod. "Yes. Sometimes I feel like we best abandon the façade and call our textbooks novels. The trouble is, if one pays too much attention to the difficulty then historians just end up collecting facts of no interest to anyone. They become antiquarians. Or stamp collectors."

The sympathy evident in her gentle smile as she echoed the old taunt of taxonomists inspired him to ask the question that had plagued him ever since Frank's visit. "Is this why you gave it all up?"

"No, though perhaps it made it easier." She paused, and then added, as though to ward off his request for an explanation, "In giving up that life I did no more than millions of men and women forced to alter their days by circumstance. I had the luxury of choice, that's all."

She picked up a set of trays and stood to place them back in the cabinet behind the sofa. Martin looked at the trays that remained on the table. He should put them away, too, and obey this apparent signal that their conversation was over, that perhaps he had pressed her too far. He should catch his boat. But he hazarded one question more.

"Do you miss it?"

"Yes," she replied. "But sometimes that isn't enough to justify going back, is it?"

He had noticed she tended to deflect questions about herself by asking him a question in return. When she said nothing more, he gave in to the voice in his head telling him "Go!" and, picking up the last set of trays, placed them in the cabinet as well and closed the doors. She had not moved away from the cabinet, which meant she was very near when she surprised him by speaking again.

"Maybe if I loved it enough, as much as I should have, I'd not have given in so easily. But I don't regret it, Martin. We ask the past to bear so much of what are in fact hopes or fears for the future. Maybe that's why people care so much about the origin of humanity. As though man is somehow one thing or the other in the present, because of where he came from. When perhaps it is better for us to believe we are really nothing more and nothing less than whatever we are in this moment. All these crusades to remake the world can make us forget that. They can make us forget the individual suffering that is near. Now."

"And so. Josiah," Martin said quietly.

She stared at his tie as she spoke. Her brow furrowed, as though she was searching for the words to make him understand.

"I know I am of use to him. I can notice the temperature of the room, and from the changing shades of his brow tell whether he is warm or cold, anxious or calm. I can determine his desires if I look closely enough. These are my facts and I can do good with them. They don't require any speculation. And I can be certain of them." Then she looked up. "You understand this, don't you?"

Martin could never decide whether his voiceless reply to her question arose from the animal or spirit in whatever he consisted. Certainly reason played no role. For he leaned forward into the space between them and kissed her.

Later he would try to make sense of what he had done, and the moments that followed. Why, after returning his kiss with a passion that unmoored his ability to think straight for days, did she suddenly lower her head and move away from him? In the wake of letting her go, he closed his eyes and pressed his forehead against the cabinet, breathing in the scent of the mahogany to try to calm his heartbeat. As his pulse slowed, reason came back with a vengeance. She read his thoughts as though they were laid out before her, spoken and explained.

"Don't be angry, Martin. It's not your fault. Perhaps there was something to be said for the old medieval cloisters. They were a stone barrier against whatever chemical reactions distract men and women so easily."

Martin turned to face her, leaning his back against the cold wood of the specimen cabinet.

"You are not a distraction," he said. "And I don't kiss every woman I stand a foot away from."

"I know that, Martin. I do. But I'm a distraction just the same."

She had crossed her arms. Was it to ward off any argument? Did she know that whatever was coursing through Martin's veins had destroyed his reticence?

"Is it Frank?"

She shook her head but said nothing. So he grasped at the specter of the other man he had turned into a rival over the past few weeks, despite all effort to reason jealousy away.

"The man who came to the police station. Charles."

"He's a friend of Frank's, and has a Ford. I didn't know what state you were in. The only information the police gave me was that you were bleeding a great deal."

Her sudden matter-of-fact manner of speaking made him realize that unless he spoke frankly she would be able to shut him out of any means of understanding why she had moved away from him.

"I must have a reason, Helen. If you don't have feelings for me, then please tell me so."

"You're a fine example of an agnostic," she said. The words were sarcastic, but her tone was gentle. "Surely you don't believe a reason can be given for everything?"

"I think we might try, if we care enough."

"Do you?" she pressed.

A sternness had entered her voice as she mercilessly used his own ways as a weapon against him. He could see she was suppressing tears.

"I'm sorry," he said, trying to convince his head or heart, perhaps both, of the words he was speaking and to accept ignorance once more. "It isn't right of me to question you like this."

Then he remembered that in a few weeks she and Josiah were supposed to join him in the field.

"You will still come to the field station?"

"Yes of course," she said quickly, and then hesitated. "If you still want us there."

"I'd be even more furious with myself if I knew my . . . if I knew this influenced Josiah's moments."

There was nothing else to do but go. As always, she followed him to the door and took his hat and coat down from the hook on the wall. Only this time, after he had taken them from her, she held out her hand, as she had done weeks ago, and said at almost a whisper, "Thank you, Martin."

Was it because she was grateful to him for stopping his questions? Or was it an unspoken acknowledgment of an understanding between them, that they continue on just as they had before? He looked at her hand and then at her faint, tired smile.

"I'll just take the words, gratefully, if you don't mind," he said. He tried to smile but he was so angry with himself that his eyes stung. He just managed to add, "Chemistry, you know."

Late that evening, Martin went to his office rather than home. His father was in Seattle for once, and he couldn't face his banter about everything that was wrong in the world. Not now. He turned on the desk lamp, took out a pad of paper and pen, and tried to focus on listing the colleagues he should contact next to find homes for Josiah's specimens. But the memory of her kiss would come upon him with a rush and the rhythm of his heart would jump. After this had happened three times in the first ten minutes, he placed the paper and pen in his desk drawer and gently closed it. Then he turned off the lamp and sat motionless in the dark in front of his stacks of specimen trays.

Chapter 15

For days Martin tried to rebel against whatever synapses in his brain interfered with his ability to concentrate on the specimens before him. The effort to read Helen's movements, with no compass but his own memory and desires, drained all confidence that he could read the patterns in the specimens on his desk. They were unrelated phenomena. He knew that. But he could not regain the focused attention that previously had kept him at a museum desk for hours on end. She had finally been willing to talk with him, to tell him something of her past, and accept him as a friend. And in the space of a moment's unthinking action he had, he was sure, destroyed all access to her thoughts and history. The only thing he had now was this maddening ignorance of why, after returning his kiss, she had halted.

He'd read plenty of explanations of human conduct that described love as an instinct. Like the desire created by hunger, this particular instinct sought easement from pain in possession. Even Darwin defined love as a feeling of anxiety due to separation. It was, he declared, a social instinct that had allowed groups to form, cohere, and outcompete other groups in the struggle for existence. Biologists were now arguing that love was a result of natural selection, and thus seeing it as anything more than a trick of brain chemistry was a dangerous delusion. If the true origins of such a behavior so vital to the future of the race was misdiagnosed, they warned, it would never be governed rightly by reason.[*]

He tried to imagine his thoughts were the result of chemical combinations so he could reorganize them by thinking about something else.

But he failed miserably in his effort to rediscipline his mind into its usual patterns and priorities, and was astonished when a line from Bryan's book kept going through his head, as the only helpful means of making sense of his feelings:

Have you ever read a scientific description of love? You never will. Why? Because a man does not know what love is until he gets into it, and then he is not scientific until he gets out again. And even if we could understand the mysterious tie that brings two hearts together from out of the multitude, and on a united life builds the home, earth's only paradise, we still would be unable to understand that larger mystery that manifests itself when a human heart reaches out and links itself to another heart.*

He was certain of one thing. He had to recover his ability to concentrate before Helen and Josiah arrived at the biological station for the summer session. Too many excellent observers would be present who could diagnose his emotional turmoil if he didn't. Especially the one who could not, in fact, see anything through physical means.

It all meant Martin almost felt relieved when Trevor asked him to warn Josiah that he might have a very distracting task on his hands for the summer.

"What do you mean?" Martin asked.

"Have you seen this?" Trevor said as he handed Martin a crumpled scrap of newspaper.

J. T. Scopes, head of the science department of the Rhea County high school, was charged with violating the recently enacted law prohibiting the teaching of evolution in the public schools of Tennessee. Prof. Scopes is being prosecuted by George W. Rappleyea, manager of the Cumberland Coal and Iron Co., who is represented in the prosecution by S. K. Hicks. The defendant will attack the new law on constitutional grounds. The case is brought as a test of the new law. The prosecution is acting under the auspices of the American Civil Liberties Union of

New York, which, it is said, has offered to defray the expenses of such litigation.*

"That's a bit odd, isn't it?" Martin said, "That the prosecution is acting under the auspices of the ACLU in a case like this?"

"Oh, the entire thing's more than a bit odd. Bryan and Darrow have both signed on to take the case."

Martin looked up. "Are you kidding?"

Trevor chuckled. "Your dad's out of town, isn't he?"

He dug through his satchel and placed a stack of clippings on Martin's desk. Martin obediently skimmed through them, reading the headlines.

SCOPES WILL FIGHT ANTI-EVOLUTION LAW; Tennessee Professor Says He Will 'Stay to the End.' Darrow and Malone Offer Aid.

SCOPES IS INDICTED IN TENNESSEE FOR TEACHING EVOLUTION; Grand Jury Acts After Judge Reads Genesis on the Creation of Man. ACTION UNDER NEW STATUTE; Schoolroom Is Declared Place to Develop Character, Not to Violate Laws. TRIAL IS SET FOR JULY 10; Science Association Plans to Aid Defense.*

"What a mess," Martin murmured.

"I didn't think you'd like it any more than I do. It's a test case, got up by the ACLU so they can get the law ruled unconstitutional. Still. A lot of damage might be done to all sides in the meantime. Newspapers are already setting it up as the latest battle between science and religion. I think you might as well warn Reverend Gray this might take some navigating, depending on how closely students are following the news. 'Course we'll be far from the papers, which is good."

Martin nodded a "Yes, sir," and thought Trevor was done. But as he picked up his hat to leave, Trevor added, "By the way, Slater says you turned down the course on eugenics."

"Word does get around fast when we shirk our duty," Martin said.

"Not at all. But you do know this is going to turn you even more Red in the eyes of some."

"I don't know," Martin answered. "I've read a few tracts by socialist eugenists. They just insist capitalism has to be abolished before women can choose rightly."*

"Did you tell Slater you'd teach the course after the revolution?"

Martin smiled. "No. And you can tell whoever cares I offered to teach a course on economic entomology instead, if they're worried."

Trevor laughed. "Depends whether you lecture about the field saving profits or the people." He turned to go with a final "Do give your minister a heads-up about this circus in Tennessee. Though we'll try and ignore it all if we can."

Martin couldn't ignore anything in the news for a week, for Will came home a few days later, stayed longer than usual, and turned the radio on at each meal. Plus Ben was having a field day. He made a big show of disgust that a trial over the foundation of biology was even possible, but he was clearly relishing the drama. He began posting newspaper notices on his office door, including a "Topic of the Times" piece that demanded to know why Tennessee didn't go whole hog and ban astronomy and geology, since they weren't based on the book of Genesis either. Another clipping appeared that claimed "the whole Dayton episode would be a subject for mirth were it not for the fact that the Nineteenth Amendment has added so many to the electorate who are intensely controlled by their emotions and sentiments." But it wasn't clear whether Ben had put that one up, or some student who thought it would be fun to irritate the coeds.*

Meanwhile Martin was trying to find new homes for the specimens required to create good maps of the relationships that evolutionists were trying to explain. The trays would have to be packed up one by one and carried to the post office, assuming Martin could find buyers through the network of collectors and curators he had nurtured over the years. Most replied to his queries with regrets. All the curators in Europe did so, often adding that they had been unable to purchase specimens since the war. There were a thousand, a million mysteries, being set aside as a result.

He kept reminding himself that he'd never had enough specimens anyway. He had, after all, spent his career trying to sort out the relationships between different forms of beetles based on very little information. Then his dad came home with a stack of files, slammed them down on the table, and said, "Hey, you're an expert in this kind of thing. Mind taking a look and telling me what you think?"

Martin was at the sink drying dishes, so he hung up the dish towel and sat down across from his dad at the table.

"What are they?"

"Bunch of birth certificates. Marriage licenses. The Duwamish are filing a claim for tribal recognition. Lawyer on the case asked me to look these over and see if I can draw up a genealogy. Here's what I've come up with."

He handed over a piece of paper with a bunch of names, with lines connecting the names, in an upside-down triangle. It was a messy attempt at a genealogical tree.

"It's all a bit of a mess, see? Most of the families are connected to several different tribes. A few upstanding English and Swedes in there too. After the treaties most had to choose one of the eight treaty tribes. Some registered as Tulalip. Or Muckleshoot. Bureau's trying to declare the Duwamish extinct. Don't have to give them schools or anything else if the campaign succeeds."

Martin picked up the piece of paper on the top of the pile. Will pulled up a chair next to him and answered questions for hours. Martin tried to draft a second genealogy, but Will couldn't give him enough information to get past most individuals' grandparents or to track brothers and sisters more than one generation back. Finally Martin suddenly halted all work, shaking his head.

"There's not enough here, Dad. The names don't match up."

"Well of course they don't match up. Half the time the whites insisted they take new names. But if we can't make a case they're related, the bureau wins."

"No. There's not enough information for that either. Surely."

"But there's enough self-interest. That'll get it done all right."

Martin watched as his father gathered up the papers again.

"Shouldn't you ask them? Rebecca's relatives, I mean."

"Sure, but their stories don't satisfy the standards of the law." Will almost spit the last two words out. "The court wants documents. A paper trail. With names, not stories. Memories aren't good enough. They want documentation." He gestured toward the pile. "And this is it."

Martin shook his head again. "I'm sorry I can't be more helpful, Dad."

"It's not your fault, Marty. The bureau's demanding they prove they're a tribe after decades of forcing them to disperse from each other and their homelands. It's criminal. And oh so damned civilized."*

After Will went to bed, Martin sat at the table for some time, staring at his dad's stack of papers. He was used to having very little information about the creatures he cared about. Faced with a gap in knowledge, he often had little or no chance of obtaining more, no matter how much he desired it. The problem was inherent to taxonomic work, even before the war had destroyed the old collecting networks. He had to wait for the luck of a package with certain specimens or rare moments in which he could comb the field himself for new specimens. By temperament or training, that fact had never bothered him. But his ability to tolerate high levels of ignorance was beginning to fail him. When he thought of all the textbooks he'd read on the better breeding of human beings. When he thought of eugenists' confident genealogies that claimed to trace moral depravity and feeblemindedness back generations. When he looked at that pile of papers, from which he could draw no certain conclusions to aid his dad's friends. And when he tried to understand Helen and reconcile himself to her wishes while completely ignorant of what she was thinking and feeling and why.

—

He had promised to warn Josiah that students might be following the trial in Tennessee but found excuses for three Fridays in a row not to make the trip. He wasn't sure how to inoculate himself against what Helen's presence did to his pulse. But eventually he ran out of excuses.

He made the trip via the interurban railway amid a late spring rainstorm that turned the landscape damp and gray. Arriving quite drenched after the walk from downtown Tacoma, he was met at the door by a stranger. A woman of a grandmotherly age eyed him a bit warily before Martin introduced himself. Her stern expression transformed into a broad smile.

"Oh good heavens! Come in, professor!" she said, whisking him into the house in a bevy of concern for his soaked trousers and muddy shoes. "Mind the boxes there. And there."

Martin glanced down the box-filled hallway as he took off his shoes, bewildered. He managed a "What's going on?"

"Didn't you know? Josiah has finally come to his senses and agreed to move in with me," she replied as she held out her hands for his socks as well.

Martin forgot about his wet socks. "With you?" he stammered.

The woman laughed. "Can't you see the family resemblance, my dear? I'm his sister, Eliza."

Martin took the question too personally and murmured a rather lame "oh" before obediently removing his socks and following her, barefoot, into the sitting room. Her hair was white, like Josiah's, but with waves of curls twisted up into a bun. She was tall and thin and moved quickly. Martin imagined Josiah had once moved like that, before he had to worry about bumping into things.

"Stand next to the radiator. Just there," she said as she laid his socks and coat on the iron bars. "It's lucky we had the heat on today. I know Josiah prefers a good fire, but I insist on modern conveniences. I suppose I'll have to put up with a fire in my parlor now. We're setting that room up as his study. I'm having bookshelves put in now."

Martin was mining his memories as she spoke, searching for any mention of a sister.

"I didn't know he was moving," he said.

"He's resisted the idea for some time. I've been pestering him about it ever since his sight began to fail. I'm in that old house all by myself. Doesn't make any sense. We'll be quite cozy, of course. Nothing wrong with that. We slept in a bedroom with six siblings as children. And what's age if not returning to one's childhood? We might as well become helpless and doddering together, for old time's sake."

Martin was too nervous about meeting Helen to ask about her place in all these new arrangements, so instead he asked Eliza where she lived.

"Quite near you, I gather. Helen tells me you live a few blocks from the university." Faced with the news Helen had talked about him, he forgot to reply, and Eliza added, "I live in Ravenna Park. I suspect your presence nearby convinced Josiah to move more than my own."

Martin was too astonished by news of the move to engage much more in conversation but Eliza didn't seem to mind. After turning over his socks on the radiator, she sat down to wait for Josiah with him and was obviously quite happy to do the talking. She had two servants, she explained, one as cook and another who came in once a week for house cleaning. She sent out her laundry. She was saying something about how lucky it was the movers would be coming while Josiah and Helen were on the island with him when Josiah finally appeared. He looked somewhat flustered amid Eliza's organized enthusiasm. When Eliza left the room to continue packing, Josiah confessed he felt quite out of sorts.

"And all I've done is sit in this chair and answer her questions all day."

"How did this happen?" Martin asked.

"She's been wanting me to do this for years. Thinks I shouldn't live on my own now that I'm eighty. Might as well keep the old girl happy."

Martin wanted me to ask why Helen didn't count but held his tongue. So instead they talked of the courses to be held at the station that summer. Josiah's queries seemed somewhat weary, and Martin feared asking his friend to serve as spiritual counselor to students in the midst of a trial over teaching evolution might be asking too much. Josiah obviously knew what was going on in Tennessee. Helen must have read some of the newspaper reports coming out over the past week to him.

"It's all quite upsetting," Josiah confessed. "The defense's plan to argue that evolution doesn't conflict with Genesis will be a hard argument to sustain with both Bryan and Darrow in the room."

"Yes. That's true," Martin replied.

His dad had read a dozen statements by scientists, writing in support of the defense of John T. Scopes. Each insisted that the purpose of science is to develop without prejudice a knowledge of the facts, the laws, and the processes of nature, while the task of religion is the development of the conscience, the ideals, and the aspirations of mankind. Properly defined, they argued, science and religion were complimentary, not in conflict. Both his father and Ben were hopping mad that the spokesmen for science were conceding so much.

"Dad's been reading a book called *The Sins of Science* by Scudder Klyce," he offered. "Klyce insists all this science-gets-nature-and-religion-gets-morality talk is merely a ploy to erect an impregnable edifice around science. That it's all just a strategy to get funds from Carnegie and Rockefeller, and shut down critics by portraying scientists as immune to human temptation and failings. He says the stance conveniently turns scientists into an infallible priesthood. And that they do, in fact, make statements about values, morality, and meaning."*

"But not all of you do," Josiah said, shaking his head. "You don't."

"No. But Klyce calls agnosticism a sugarcoated name for incompetence. It's a highbrow euphemism for quitting. Morally futile. And he says any man who turns his money over to agnostic professors is paying for the destruction of his children."

"Sounds like a fine ally for Bryan," Josiah said.

"That's the funny thing. Klyce is no fundamentalist, and no friend of Bryan." Martin did not add as explanation that Klyce described religious experience as equivalent to an orgasm and lamented prohibition on the grounds a little ritual intoxication might revive the church. Instead, he explained that Klyce completely denied the divinity of Christ, original sin, and the fall of man.

"He believes a God who would separate us from our loved ones by routing individuals toward heaven or hell is a monster to whom no man in his right mind would give allegiance. Klyce is, on these counts, everything Bryan fears. Still. Their target's the same. And Klyce isn't just insisting taxpayers stop paying for their children to be taught the evolution of man. He says Americans shouldn't pay the salaries of scientists at all."

"Good grief," Josiah said. "Sounds like you're being pummeled from all sides."

"Yes," Martin said with a smile. "It all makes a summer in the field look quite appealing. If we can avoid the newspapers."

Eliza descended on the room to give Josiah a complete update on what she'd packed so far and to question him on about a dozen items she wasn't quite sure needed to be moved to Seattle, given they'd be short on space. Josiah replied with evident fatigue, while Eliza looked somewhat harassed, as much as she seemed to enjoy organizing things. When she left the room, Martin stood up.

"This is quite an operation. What can I do to help?"

Josiah gave a tired laugh. "I haven't any idea. You can ask Eliza, if you can bear a dozen orders. Or you can ask Helen. Yes of course. Ask Helen."

"All right," Martin said, though his heart beat quicker. "Where is she?"

"Packing her things I think. First door to the right, upstairs."

Martin had never been on the second floor. When he came to the open door to Helen's room, he tried to imagine what, a few months ago, he would have thought if met by the floor to ceiling shelves within, packed tightly with neat rows of books. Helen was sitting on the bed with a few books in hand and boxes strewn about her on the floor. He had steeled himself against this moment and spoke as casually as he could.

"Hello. Josiah sent me up. Can I help?"

Helen looked up with neither a flinch nor a blush. As though it was quite normal for him to suddenly appear at the entrance to her bedroom. Martin received no indication that she had suffered during his absence. Or that she had tried to forget the look in his eyes when he gently refused to shake her hand. Or that she was calling on well-trained reserves of stoicism at this moment.

"Hello, professor," she said. "I suppose you can help with a little extractive surgery on my mind."

Her words as she placed the books in her hand in a box knocked Martin's nervous anticipation of this first meeting from his mind.

"What do you mean?"

She picked out a few more books from the shelf and looked at the bindings.

"There isn't room for these in Seattle."

She placed the books in a box, and Martin realized what she was doing.

"No!" Martin said the word without thinking. Softly. Under his breath. Maybe she didn't even hear him. But then he spoke louder. "Does Josiah know?"

"It isn't worth worrying him about," she replied. "I can't in good conscience make a case to Eliza that I need them anymore. She, quite rightly, wonders why I have them."

At that moment he knew how she had answered Lucy Salmon's letter urging her to return to Vassar. He started to speak, but she looked up at him and spoke quickly.

"It's all right," she said, and then added a gentle but firm command. "Please don't argue." That was when he saw the emotion flash across her face. As though she was afraid of his arguments.

Martin stood very still, looking at the boxes.

"What are you going to do with them?" he asked finally.

"Take them to the bookstore downtown, I guess." And then she looked up at him again. She had recovered her composure. "There is something you could do. Could you take them? I'm not sure I'd have the strength to stand there as the owner tallies the value of each volume. I can pay you for the trouble."

"Only if you promise not to speak of payment of any kind."

She thanked him and returned to her task. He couldn't see her face anymore. Although he knew her quip about helping with surgery on her mind had not been a serious request for aid, he entered the room and knelt down by one of the boxes.

"If it's all the same to you," he said looking over the titles of one stack, "I'll take them to one of the bookstores in Seattle. They'll give you a better price than anyone in Tacoma."

"If you like," she answered. He thought they would work in silence, but after a moment she added, "I doubt any of these will draw a high price anywhere." She smiled slightly as he looked at her. "There aren't many who want to read about an Age of Faith in an Age of Science." She paused, and then spoke again. "Somehow I keep ending up in your debt, Martin."

He tried to smiled in return. "I figure asking a humanist to read almost seven hundred pages of *The Descent of Man* aloud should cost something." He wanted to add: But not this. Not helping her dispense with an entire library, one she obviously loved a great deal. He asked instead, "Shall I read the titles aloud just in case you wish to keep a few?"

"I have chosen a dozen," she replied. "You are not near them."

Then they worked in silence. Some months ago he had gone to the university's library, struggling to reclassify Helen in his thoughts. Now he sat, surrounded by evidence of her history, and forced to place it all in boxes to be sealed up and sent away forever.

Eventually Eliza brought them tea. Martin thought they would take a break, but Helen kept packing. He picked a book he had set aside and sat back against some of the shelves, stretching his legs out in front of him. It was a medieval bestiary, all in Latin. She had, of course, understood every risqué line in *The Descent of Man* better than he did. Still, his years of dealing with long Linnaean names served him well. He could even handle the untranslated passage from the book of Job on the frontispiece.

> But ask the animals, and they will teach you,
> or the birds of the air, and they will tell you;
> or speak to the earth, and it will teach you,
> or let the fish of the sea inform you.
> Which of all these does not know
> that the hand of the Lord has done this?
> In his hand is the life of every creature

and the breath of all mankind.
(Job 12:7–10).

As he turned to the medieval stories of animals, at least once a sentence he asked her for the translation of a particularly ecclesiastical word. He had read a few entries, and decided to read one more, when he asked her for another translation and she replied, "Firestones. Magnetic rock." The blood rushed to his face as he read the rest of the entry in silence, given she now knew what he was reading:

> There are fire-stones, in a certain oriental mountain: they are male and female. When these are far apart from one another the fire in them does not catch light. But when by chance the female may have approached the male, at once the flames burst out, so much so that everything around that mountain blazes. You, therefore, O Men of God, who wage this way of life of ours, keep well away from women, lest, when you have got near to one another, the twin-born flame shall break out in yourselves and burn up the good things which Christ has conferred upon you.[*]

He closed the book and placed it in the box.

He wished he could see the ones she had set aside. As facts by which he could judge what she valued most. The titles on the spines of books in German, Italian, Latin, and French were constant, painful reminders of how poorly he had classified her for so long. Then he came to the *Life and Letters* of the agnostic biologist Thomas Henry Huxley. This book was in English, and meant a great deal to him.

"My father read me this when I was ten," he said, to explain why he paused again. He held the spine up so she could read it.

She smiled. "I could have guessed that."

The book's binding was quite worn at one spot. He let it fall open naturally in his hand, and read the opening lines of a letter he knew well, one composed after Huxley's son Noel had died at the age of four. Rev. Charles

Kingsley had urged Huxley to take comfort in the belief he would meet his son in heaven. Martin silently read Huxley's familiar reply:

> Had I lived a couple of centuries earlier I could have fancied a devil scoffing at me—and asking me what profit it was to have stripped myself of the hopes and consolations of the mass of mankind? To which my only reply was and is—Oh devil! Truth is better than much profit. I have searched over the grounds of my belief, and if wife and child and name and fame were all to be lost to me one after the other as the penalty, still I will not lie.*

Kingsley had asked Huxley to believe in immortality and be consoled. But Huxley demanded whether it was right to believe in something because he liked it. Science, he explained, had taught him the opposite lesson:

> She warns me to be careful how I adopt a view which jumps with my preconceptions, and to require stronger evidence for such belief than for one to which I was previously hostile. My business is to teach my aspirations to conform themselves to fact, not to try and make facts harmonise with my aspirations. Science seems to me to teach in the highest and strongest manner the great truth which is embodied in the Christian conception of entire surrender to the will of God. Sit down before fact as a little child, be prepared to give up every preconceived notion, follow humbly wherever and to whatever abysses nature leads, or you shall learn nothing.

Martin thought of Helen's words, before he had disobeyed his deep habit of self-control and kissed her, that he reminded her of the medieval theologians who distrusted their ability to avoid self-delusion and discern God's purposes and ways. Perhaps Sarah's minister-friend was right, he thought. And they weren't so different after all.

He closed the book and gently turned it over in his hands as he spoke. "Our copy fell apart. Dad read this letter to Reverend Kingsley so many times after my mother died. We didn't have a service. He just read this letter." He paused, before adding, as though to himself, "Strange I remember that."

As he placed the book in a box he realized the sound of her quiet movements had stopped and he looked up. She was watching him, so he saw the unguarded tears in her eyes. Before he could voice the "What is it?" that caught in his throat, Eliza appeared in the doorway to deliver a lengthy update on her negotiations with the movers as she collected their teacups.

Martin had been inclined to like Eliza well enough. Now, given her apparent indifference to what Helen was doing, he had a hard time forming a civil reply, especially when she expressed delight that Martin would be taking care of the boxes of books stacked in Helen's room so that they need not find a place for them when the movers arrived. She insisted on paying him for the trouble of adding some of Josiah's most valuable boxes to the stack.

"A young college professor cannot afford to be so generous," she teased. "Not if he wants to catch a wife before he's sixty."

Hours later, Martin was on the porch pulling up his collar in preparation for confronting the rain again, when he heard the door behind him, and turned to see Helen. He had not been alone with her since Eliza had descended on her room, for this time it was Eliza who had walked him to the door and retrieved his hat and coat.

"Since you won't let me pay you for your troubles will you take this instead?"

She handed him a book and Martin looked at the familiar spine. She anticipated the no on his lips and said quickly, "Please take it. I know much of it by heart."

He obediently tucked the book in his satchel.

"All right. I'll try and get a good price for the rest."

She nodded in acknowledgment. There was nothing else for him to do but step out into the rain. To say "Are you sure?" would have amounted to a painful challenge. He could see that well enough.

When, soaked and cold, he found a seat on the boat to Seattle, he opened Huxley's book to where it naturally fell open once more. Had Helen broken the binding here? Had the book been Frank's? She said she knew much of it by heart. In the midst of Huxley's explanation of why he had abandoned belief in immortality, a single sentence was marked with a pencil. *Love opened up to me a view of the sanctity of human nature, and impressed me with a deep sense of responsibility.* He read the lines that followed:

> If at this moment I am not a worn-out, debauched, useless carcass of a man, if it has been or will be my fate to advance the cause of science, if I feel that I have a shadow of a claim on the love of those about me, if in the supreme moment when I looked down into my boy's grave my sorrow was full of submission and without bitterness, it is because these agencies have worked upon me, and not because I have ever cared whether my poor personality shall remain distinct forever from the All from whence it came and whither it goes.

The memory of how she had looked into his eyes rather than try to hide the effect of his speaking of this letter made him close the book and set it beside him on the bench. Bryan was insisting Darwin's theory would chase love from the earth. But Martin was certain that no scientific explanation could banish what he had felt at that moment. With a rush of emotion he rested his head in his hands. Bryan contrasted the supposedly stark determinism of evolution with Christianity, for only Christianity could change a heart in the twinkling of an eye. "When one's heart is changed," Bryan wrote, "when he is born again—he listens to, understands and accepts arguments that he rejected before." Martin had never envied Bryan's certainty about the capacity of human beings to rightly figure out what they should believe and do, so long as they had faith in God. But he did now. Faced with the demand that he both forsake anything more than friendship and live in ignorance of the reasons, he envied Bryan's faith in the ability of a struggling individual to transform and reform his desires, so long as he called on God's aid. He was not sure, left to his own devices, that he could do it.*

Chapter 16

"The Science Service is readying the troops for Dayton," Ben announced, tossing a copy of the *Science Newsletter* onto Trevor's desk. Trevor groaned and moved the newsletter off his six-inch-high to-do pile. Martin and Pete were in his office to help pack a brand-new set of microscopes for the station when Ben appeared with his news. Martin picked up the newsletter and read Osborn and Davenport's call to America's fourteen thousand scientists to support freedom of teaching. The call included a resolution issued by the Council of the American Association for the Advancement of Science that Darwin's theory was "one of the most potent of the great influences for good in human experience."*

"They're having a lot of trouble drumming up expert witnesses for the defense team," Ben continued, ignoring Trevor's lack of enthusiasm. "Most men'll be in Europe for the international congresses. Metcalf and Curtis are going though, and Kirtley Mather. Apparently when they asked Morgan for help he just laughed and shrugged his shoulders. I don't know who he thinks is going to pay for all his flies and milk bottles when we're being accused of teaching antisocial ideas."*

"I suspect Morgan has better things to do with his time," Trevor replied a bit too defensively. "This trial's going to be a fine platform for Bryan's speeches on how Darwinism is destroying all love, sympathy, and humanity. And that we're to blame. If the citizens of the state of Tennessee want to keep their children ignorant, there's no need for good scientists to waste breath trying to enlighten 'em."

"Bryan gets newspaper space well outside of Tennessee," Ben countered. "And surely we want educated rather than ignorant citizens! Besides, one never knows where the next Darwin might be born."

Trevor laughed. "Well, that's hardly going to convince those who believe Darwin is responsible for all that's wrong with the world. I hope the defense has a better strategy for defending this man Scopes."

"Whatever line the defense takes, we'd best step up and do our part," Ben replied. "I've heard William Beebe has got the funding for an expedition to the Galapagos to settle the matter once and for all. Get a good collection of finches and lay their diversity out before the public. Let them see the evidence for themselves."*

Ben noticed Martin's furrowed brow. Lately he'd lost his ability to maintain a good poker face in response to Ben's speeches.

"Well, Sullivan?"

"I don't think all the Galapagos finches in the world are going to settle this. Seems everything's moved a bit beyond finch beaks."

"As it damn well should," Ben answered.

Trevor changed the subject back to how many microscopes they needed to replace. After Ben got bored and left, Trevor confessed to Martin that he hoped to keep debates at the fireside about God or what animals taught anyone about human nature to a minimum.

"Glad we're leaving before this circus in Tennessee starts," he said.

"Dad says the trial's going to be broadcast by radio," Martin replied.

"Well. There aren't any radios at the station," Trevor said, pulling hard on a knot he'd made in the twine around one of the microscope boxes.

—

Sitting in his office a few days before the trip to the biological station, Martin had opened another letter from a museum curator in Sweden expressing regret that the national museum in Stockholm possessed no funds to purchase specimens. He began typing another query, this time to an entomologist in Spain, when a knock sounded on the wall alongside the open door. He glanced up a bit absently, having finally regained his concentration, and immediately lost it again when he saw Frank Gray standing in his doorway.

"Still trying to sort out the world?" Frank said.

"Just a little piece of it," Martin replied, yanking the paper out of the typewriter.

He had pressed a wrong key. He felt a pang of regret for speaking so shortly. Though Frank was connected to the loss of the collection, Martin agreed with Trevor's suspicion that the funding for a natural history museum probably would have been lost to other priorities soon enough. But he was tired of trying to concentrate and overwhelmed with the task of finding new homes for thousands of specimens. And here was Helen's husband standing before him. He motioned for Frank to sit down.

"What are you doing here?" he asked.

Frank ignored the chair and leaned over to peer beneath the glass of the nearest specimen tray.

"Back from my field trips and on my way to Chicago. I leave tomorrow. You teetotaling today or up for a drink?"

Martin looked at his paperless typewriter, the long list of correspondents, and the trays to be packed alongside it. He stood to get his jacket.

He fell into step beside Frank, who obviously had someplace in mind they could find a drink at midday. Frank seemed deep in thought as they walked west toward University Avenue. Once they were sitting down in the basement of a "laundromat" with a scratchy recording of a George Olson fox-trot in their ears, Frank ordered and placed unidentifiable drinks in front of both of them. He downed his quickly, while Martin took a mouthful and coughed. The drink was poor even by prohibition standards. Frank waited for him to recover and then leveled a steady gaze on him.

"All right, professor. What's this I hear about you helping her get rid of her books?"

Martin stared at him blankly.

"I just came from Eliza's," Frank continued. "She thinks you're quite the gentleman for helping them with the move. Why'd you let her do it?"

"I didn't let her do anything," Martin replied firmly. "When I asked if I could help, that is what she asked me to do."

Frank shook his head. "A man can say no. You could have said no."

Martin didn't trust himself to reply calmly to that, so he said nothing.

"Did you know she turned down the position at Vassar?" Frank pressed.

He spoke with an almost angry, accusing look in his eye, as though Martin was complicit in that fact, too. Because he had carried away those boxes of books. Martin knew then that Frank had intended him to read Professor Salmon's letter. But he rebelled from playing a role in some scheme to influence Helen's decisions. He tried to redirect the conversation away from her entirely.

"I thought you despised the ivory tower."

"Look, I wouldn't wish that world on Helen," Frank replied. "But it would be better than how she's spending her days now."

Martin gave into Frank's focus on Helen's moments with undisguised exasperation in his voice.

"Why do you begrudge Josiah her attention? What's the point of a campaign for equality if you still get to say how she should spend her days?"

"Ah, Sullivan. Don't bring ideology into this!"

"I thought it was a language you might understand. Or would you prefer some psychology? Maybe you're just jealous of your father?"

"Well now, that would indeed be some nice fodder for a Freudian," Frank muttered. He sat back in his chair, staring at his drink.

When after some moments he looked up again and spoke, the note of accusation had died away and he just sounded tired.

"That's a pretty good scar."

"Looks dramatic, I know," Martin replied, resolving to match Frank's changed tone, so long as he didn't mention Helen again. He smiled slightly. "I didn't really earn it though, did I?"

Frank smiled in return.

"No. You didn't." He fell silent again for a moment, and then took a deep breath. "We hadn't been raided in three years. I wouldn't have dragged you along if I'd suspected they were still paying us any mind. Seems the police chief was just bored. Damned fellow let us all out at three in the morning. Raid didn't even make the newspapers. But you gave me a damned good fright, when I saw you keeled over against the bookshelf covered in blood."

"I don't quite understand how you got me out."

"Helen's book and a bunch of entomology reprints. Hard to argue with my claim you weren't one of us after finding those in your satchel." He paused a moment, and then took a long drink before adding, "Whole fiasco meant I never got to ask you what you thought of my speech. Too hard on you lot?"

"I agreed with much of it. Wholeheartedly. The parts I can remember. But you are on the precipice of anti-rationalism. I don't blame you, given how science has been used to justify all kinds of nonsense. But surely the ideal is still useful. In training a man to mercilessly test his beliefs."*

"And how good are your own men at this demand?"

"Absolutely miserable half the time. But you said something about doing an experiment to test whether feeblemindedness is due to heredity or environment, didn't you? Surely that means you do need us. For the way of thinking. For the methods."

"I'll tell you what most men do who take your stance seriously, professor. They withdraw from political life altogether, stifled by doubt and uncertainty. In the interests of cries for more facts, more investigation, action is stopped. There's an anthropologist at Berkeley who used to write reams for the radical press condemning the view inferior races are transitional between apes and humans. Now he spends his days doing research from a plush office in, yes, an ivory tower. You wouldn't know he even had political views."*

"I bet Madison Grant would disagree with you on that point," Martin countered. "Some of those facts and investigations are precisely what's needed to undermine all these racialist ladders. One of Seattle's anthropologists spent years measuring the heads of the children of Japanese immigrants. He's shown they can change in a generation. And that means all these hierarchies based on the cephalic index have no value."*

"But that's not one of your men. What about zoologists? What about the ones classifying butterflies like neuter pedants or chasing down bark beetles so the lumber barons can improve their profit margins?"

Martin smiled at the jibes. They cast him as at best a useless dilettante, and at worst complicit in the suffering caused by industrial capitalism.

"Know how the Soviets are dealing with naturalists since the revolution?" Frank continued. "No more private collections built on the backs of the proletariat. Science must be of some good to the people, rather than serving the whims of the moneyed classes or some man's idle curiosity."*

Martin wished he could toss Trevor's work for the oyster industry at Frank, but he doubted that was the kind of good-for-the-people work Frank had in mind. And Martin had spent the past year getting a collection of dead insects ready for a new natural history museum that, until very recently, was to be named after one of the wealthiest timber barons in the world.

Instead, he asked, "Don't you find it a bit hard, though, determining what's good for the people?"

"Most difficult thing in the world," Frank replied. "And then, once you've decided, you've got to classify human beings according to whether they agree with you or not. Those are the shoals upon which the revolution in Russia is being destroyed. Men and women who were united in 1917 splintered into factions. Anarchists. Bolshevists. Mensheviks. Then they started killing each other. Those who weren't already starving to death."

Martin could only imagine the scenes passing through Frank's mind as he spoke. His father had read him countless passages from anarchist Emma Goldman's account of postrevolutionary Russia. She called the book *My Disillusionment in Russia*. He had been present when Ben found the book in the hands of a student and ridiculed Goldman's view of human nature. "That's what comes of adopting some fool view that human nature's rooted in mutual aid and cooperation!" Ben had exclaimed. Frank hadn't appealed to Russia's example during his speech in Tacoma. He hadn't even mentioned the coming, inevitable revolution in America, usually obligatory oratory for Wobblies. Martin could become immobilized by the complexity of a group of insects, but at least he could take refuge in the task of amassing more specimens. The complexities facing Frank staggered him. The man obviously felt, rather than ignored, them. Martin did not begrudge him the call for another round of drinks.

"Well now, Sullivan. Let's see how good a scientist you are at observing your own thoughts, and testing your own delusions," Frank said, handing him another glass. "You said before that inhibitions can be useful. But I don't really trust a man to tell me the truth about himself until he's under some consciousness-numbing intoxicant. Or he's in the midst of a delirious fever that shakes him from all his civilized mores. And there's nobody here to hurt but each other if we get out of hand."*

Martin figured the fog in his brain these past few weeks couldn't be made much worse, even by bootlegged alcohol, and took the second glass.

"I don't generally have enough wits about me to observe anything well when I drink."

"Ah, but a man must drift from his moorings to really know himself," Frank replied. "Surely you'd agree your wits get in the way of observing rightly? Alfred Russel Wallace thought of natural selection during a bout of malaria. La Mettrie discovered that man is a machine, nothing more, when under the influence of a feverish delirium."

Martin tried to down another swallow. This time his throat did not rebel, and he managed a "Who?"

"My dear fellow," Frank answered. "Men think it was Darwin who destroyed God and the soul by materializing mankind. But a physician named La Mettrie did it first. Almost two hundred years ago."

Frank took a long drink after finishing his little history lesson.

"And you think we're his natural heirs?" Martin asked.

"Not at all. You lot use the materialism as a method while piously insisting you say nothing about God. No. It's men like Clarence Darrow who carry La Mettrie's banner, who try to save us from ourselves without God's help. Did you know it was La Mettrie who first diagnosed criminality as a physiological illness rather than moral failing? Of course he had to abandon Genesis to do it. He needed a new history of man to insist that the socially maladjusted among us deserved pity rather than hatred. The machine can't be at fault if he just happens to be mismade."*

"Surely such calls for compassion were taught by Christ as well," Martin countered.

"Yes. But there are quite a few men, decrees, doctrines, and papal bulls since the man of Galilee, aren't there. The only way to get rid of all these corrupt doctrines is strict materialism. La Mettrie understood that. Clarence Darrow understands it."

"But it's strict materialism that gets you to the likes of Mencken and those who claim a man is what he is solely because of heredity."

"Strict materialism and stupidity," Frank replied. "La Mettrie didn't think that through materialism he could remake men by better breeding. He only thought one might understand them better. And that a better understanding of human nature might lead us to be a little more tolerant of our fellow man, mismade machine or no. Maybe even improve their environment and education, so they could reach their fullest capacity. Without the need for a single prayer or minister's approval."

Frank punctuated his final sentence with another gulp of his drink. Martin had finished just one to Frank's three, but his system wasn't as used to the effects. Perhaps that was why he asked an unguarded question.

"And did this French doctor extend his materialism to love?"

Frank slowly smiled, leaning back in his chair. "He'd have to, wouldn't he? Whence the extraordinary interplay between certain muscles and the imagination?" Frank winked at him. "Ah, but that's not what you meant is it? He'd have to say love is nothing more than a manifestation of some physical reaction, right? If he's consistent? Well, then, what of it? Is love's beauty weakened because a scientist explains it as a complex chemical reorganization of the brain, in which images of one's beloved dominate over others? Must we renounce love as unreal because it's rooted in our animal past? Is its power destroyed? No! The patient feels it. The feeling influences his digestion. The good physician must look at the thought, take it seriously, and counsel the patient's mind and heart even as he dispenses a prescription for his bowels."*

Martin was concentrating too hard on following Frank's logic to form a reply, and startled when Frank suddenly leaned forward and whispered, "How's your digestion, professor?"

"I have a hard enough time studying beetle anatomy," Martin replied. "I suspect my knowledge of my own system isn't very trustworthy."

"Ah, Sullivan," Frank said, sitting back in his chair again. "You'd have a great deal to say, if only you'd come out from behind your museum cabinets." He slammed down another empty glass. "Speaking of which, I hear you lost the museum. And I'm sorry to say about a half hour ago I thought you deserved the loss."

"You've got some pretty good informants."

"You've no idea. Do Helen and Josiah know why?"

"No. Do you?"

"Apparently there was a round of questions asked about you. Mini-investigation, if you like. I'm sorry about that. I'm sorry about the whole damn night."

Martin said nothing, staring at his drink. He appreciated the apology, though, and that Frank could regret another man's loss despite a lack of sympathy with the actual work. After the quick downing of another drink, Frank's voice changed.

"Now then, Sullivan. I'm not sure I approve of this. Her going on this field trip with you."

Martin looked up. Frank's voice was stern, but Martin couldn't quite tell if he was serious.

"So that's why you're here? You don't care about the books."

"I care about the books because I care about her. Don't tell me you'd blame a man for keeping an eye on his wife."

Martin was tempted to down his drink and call for another one as imaginary scenes tumbled through his mind. He'd already learned Frank must have gone home again. Now he wondered whether there had been a scene, and Frank had accused Helen of something more than abandoning her work.

"You consider yourself still married?"

Frank raised his eyebrows.

"Quite the radical now, aren't we? Careful, professor. Word of thinking like that gets back to your university president and he'll have you out on the street in a minute." But after a pause and another swallow, he answered Martin's question. "Legally we're still married, yes. But I take it you appeal to something else. Well, since you asked, I don't in fact think we are. But that doesn't absolve me of certain rights that I demand as her friend."

This time it was Frank who held Martin's gaze. Perhaps it was an acknowledgment of Frank's candidness and his obvious concern for Helen, as misdirected as Martin thought that concern. Perhaps it was the alcohol, drifting him from his moorings. Or perhaps it was the grave attention of the man across from him. For Martin said something he had not expected to divulge to anyone, much less Helen's husband.

"If it's any consolation, she won't have me."

Frank looked at Martin for a long moment.

"Did she offer a reason?" he said finally. "She told you it was Josiah, didn't she? That she can't abandon him. She wouldn't say I'm the trouble."

Frank took Martin's silence as assent. "She's lying, you know."

Martin quickly regained his ability to speak. "I won't have you telling me what she is thinking."

"Well she won't tell you. Did she? Did she tell you it's because the doctors told her she isn't fit to have children?"

Martin started to stand in an involuntary effort to escape from Frank's explanation.

"No, listen!" Frank reached out and grabbed Martin's hand as he spoke. "Did she tell you we had a little boy?"

Martin sat down again, as though pushed back into his seat by the words. Frank must have read an answer in his movement.

"His name was David. He was never strong, poor little tyke. He got sick during the second wave of the flu epidemic, and then just wasted away over time." Frank paused, and then spoke with difficulty. "It took a year."

Martin felt sick, but he couldn't bring himself to stop Frank from speaking. Frank's words began coming at a faster clip through a combination of emotion, memory, and liquor.

"We both went a little mad, I think. I couldn't stand her grief. It was silent. Silent. The doctors told me to send her away, that the loss had affected her mind. They said her soundless despair wasn't normal, and I agreed to avoid going mad myself. Dad did everything I should have done, when I was away. He had seen what I couldn't, blind as he is. I handed her over to medical men who insisted they knew the material cause, when in fact they know nothing of the human mind or heart! They said she had hereditary neurasthenia exacerbated by intellectual work. How's that for a scientific diagnosis? They said she'd drained her vital energy in cultivating her mind as though she were a man. Do you understand? They convinced her our boy never had a chance in the first place. Because of how she had lived."*

Martin felt as if the content of his glass was returning to his throat. Then Frank added mercilessly, "You know where all this comes from."

Martin did know. His stack of books on eugenics contained all sorts of psychologists' theories that women who pursued education beyond home economics misdirected their finite supply of vital energy from their reproductive organs to their brains. They cited data on higher infertility and infant mortality rates among the children of college graduates as proof, and argued that campaigns to educate women outside their natural sphere would undermine all eugenic reforms and result in the degeneration of the race.*

He swallowed hard and finally forced himself to speak. "There are good scientists who would call that a load of nonsense."

"No doubt. And they're deemed superstitious and ignorant of how nature operates. That's what they called Dad when he fought the doctors for custody of her and denied their mad little explanations. He can't forgive me. Says I should have known better than anyone that they would destroy her with their fairy tales. She's abandoned her work. And spends her days like some damned Benedictine nun."

Martin leaned forward and placed his head in his hands in a vain effort to ward off Frank's version of Helen's past. But he could not halt Frank's next demand.

"Well, professor, what are you going to do?"

Martin shrugged in return, though every emotion within him rebelled against the gesture of indifference. Much more than ideology had been behind Frank's tirades against giving too much power to scientists. Behind Josiah's silence on some sections of *The Descent of Man*. That realization made the movement feel irresponsible and heartless. But he could imagine no other answer than the one he gave.

"Nothing."

"But you know those doctors are wrong!" Frank said.

That made Martin angry, and he betrayed it in his voice.

"How can you imagine that what I should do is clear? I would be speaking as more than a disinterested observer in the case! And no matter what I feel, I'm not going to insult her by argument. Especially against an explanation she did not give."

Frank watched as Martin picked up his glass so his hands had something to hold.

"I only told you because I thought you're man enough not be frightened away from her by it," he said with a strange mix of sympathy and exasperation. "Perhaps I was wrong."

Martin slammed his glass down.

"Good God, Frank, the confidence with which you judge men! I can tell you I'm man enough not to care a whit what you think. I aim to be guided by Helen's reasons and desires, not yours. Not even my own."

Frank was silent for a moment, staring at Martin's glass.

"These fundamentalist ministers are insisting that evolution leads to carnality, Bolshevism, and the Red flag. But you're more Puritan than the lot of them."*

Martin took a deep breath before replying. "I suspect you are too."

"Well, I'm not supposed to have a sense of obligation to anything outside my own selfish desires. No sense of duty or clear idea of right or wrong. Tempts a man to counter expectations, just to spite the belief that atheism leads to immorality."*

Martin grasped at the one question, of the dozen running through his head, that he could voice quickly and simply.

"Why did you tell me this?"

"Guilt. Love. Revenge on fate. I don't know. I thought as a biologist and a friend, you might be able to help her. When Eliza told me about the books I was furious with myself because I thought I'd read you wrong."

"Why can't you?" Martin said. "Why can't you help her? You're her husband."

"I told you we aren't married. Not in spirit anyway, or whatever you want to call it." He began quietly tapping his empty glass on the table. "Never was a true marriage. I mean, we had the certificate and everything. But she married me to keep me out of prison."

"I didn't know having a wife could protect a man from jail."

"Can be a pretty good distraction if he's suspected of having unnatural passions toward his own sex. Aren't many things worse than a crime against nature, is there?" Martin saw Frank shrug for the first time. "You see, professor, I'm one of those broken machines, according to your biologists. Something in my moral compass is degenerate." He smiled, and then became quite serious. "Law insists I'm a threat to public safety and must be either reformed or locked up. And biologists say I'm a congenital misfit. There are judges in Oregon who'd have me sterilized in the name of preventing race suicide." He paused for a moment, until Martin looked up. "Those laws can put a man who creates trouble for capitalists in jail for a long time, whether they can get him on a crime against private property or not. Helen's dad was the district judge in charge of my case. We were old family friends, see, but had, shall we say, ideological differences. He was after me. She saved me. Knew her dad would want the world to think she'd made a respectable match. No matter how many speeches I gave about the war or revolution. It's kept me out of jail, and worse." He gave a quiet laugh. "So. Want to take that accusation that I'm a puritan back?"*

Martin was becoming exhausted by the new information he was having to process. Yet he hazarded speaking one question more, though every new fact created a dozen new mysteries.

"But you had a child."

Frank smiled again. "Living things can't be pigeonholed so easily. You know that, professor."

Martin sat back in his chair. Perhaps Frank could tell he had heard enough, for after taking some coins from his pocket and placing them on the table to pay for the drinks, he stood up. Martin watched him as he took his jacket over his arm and his hat in hand.

"I'd have given her a divorce so she could marry you."

"You should give her a divorce whether she wishes to marry me or not." He looked up, so that Frank did not misunderstand him. "If she wants one."

Frank put on his hat.

"You're a good old conscience, professor. But remember. None of us are on our own in this. You lot have claimed to know us better than ourselves. Because you know where we came from. Comes with a certain amount of responsibility, I think."

He held out his hand. Martin stood as well and gripped the offered hand for a moment. Then Frank was gone.

He could barely think straight when he finally left the speakeasy. It wasn't the alcohol. Not entirely. Frank's explanation had connected Helen's movement away from him to biologists' campaign to influence how individuals should think about themselves and the grounds on which decisions should be made. That realization finally destroyed his ability to disconnect all his taxonomic work from the troubles of his own kind. Words that Darwin had used to transition from birds and monkeys to human mate choice passed mercilessly through his mind as he walked: "The final aim of all love intrigues, be they comic or tragic, is really of more importance than all other ends in human life. What it all turns upon is nothing less than the composition of the next generation. It is not the weal or woe of any one individual, but that of the human race to come, which is here at stake."*

Martin knew that if Frank had diagnosed Helen's thinking correctly there were biologists who would insist there was nothing irrational about her decision to end their kiss. They would even call her judgment ethical. Evolution, they would say, had revealed the need for a new eugenic religion and new doctrines of self-sacrifice: the duties of parenthood for the healthy, and abstinence from marriage if mental or physical defects were present. Nowhere could Martin remember reading, amid these calls to create a heaven on earth, any acknowledgment of the new modes of heartache that biology's supposed moral imperative might create. Yet, if Frank was right, Helen was suffering now because of what some biologists believed. If, that is, the very tangible, objective fact that she had kissed him in return was evidence she cared for him. And Martin had no way of knowing whether that interpretation of what had happened between them was correct.

He walked home under a clear blue sky, but for once he did not notice the mountains in the distance. He had never been so overwhelmed by the sense he must do something, and the drive bewildered him. He had no idea how to harness it into right action. He wished his father was home. A man with so much certainty about what was right or wrong with the world might be able to tell him what to do. But Will might not come home for weeks.

Martin paced the apartment twice. Then he sat down in front of a pile of carefully stacked boxes sitting on the floor of his bedroom. He opened one of the boxes and set the contents in smaller stacks against the wall so he could see the titles on the bindings. He began sorting the books as he stacked, guessing at their classification based on the five or six clues on the title page, and placing authors together when he saw a pattern in their names. When he was finished with three boxes, he sat back and looked at the result. The act of sorting once he had established firm criteria had set his mind at ease for a time. Then, as he sat looking at Helen's neatly organized books, Frank's words came back with a rush. So he did something he had only done once in his life. He got up, took down a medicinal bottle of whiskey kept in a cabinet for emergencies (he was sure his father would agree this counted), and drank a quite nonmedical amount.

Part V

Trials and Tribulations

Chapter 17

"I'd like to have a word with you, if I may."

At least, Martin thought that was what she said, after Mrs. Macleod knocked on the door of the apartment early the next morning.

"Downstairs."

Martin couldn't determine whether he imagined the stern tone. He couldn't quite tell whether he was standing upright either. His head felt like it was in a vice that tightened with every movement. He made it downstairs, grateful that he hadn't bothered to undress the night before, so he need not figure out how to button his shirt or trousers now. Somehow he navigated his way to the parlor, where he desperately wished the radio wasn't playing so loudly in the corner. Why did she have the room shut up on such a warm day? He gratefully obeyed Mrs. Macleod's request that he sit down, and then focused on disobeying the overwhelming desire to drop his head into his hands so the room would stop spinning as she talked.

"You're a decent young man, Professor Sullivan. I hate a scene as much as anyone, but this can't be helped. Your father. He simply must go."

Martin must have mumbled a bemused query, for she launched into a lengthy defense. The sharp jangling of her bracelets punctuated each point. She ran a respectable household. She had upstanding neighbors. This was a good neighborhood. She had tried to ignore the rumors that his father was a communist. He seemed like such a gentleman. But she could not turn a blind eye any longer. Not to such flagrant indecencies.

The neighborhood wouldn't stand for such comings and goings, nor would she. If a man wanted to live downtown, or on an Indian reservation, where such behavior was tolerated, then he should move. He must move. This week, if possible.

"Who's moving?" came a cheerful voice, and discerning that Mrs. Macleod's attention was directed toward the door, Martin rested his head in his hands.

"You, sir," Mrs. Macleod said firmly.

Then there was something about a "half-breed" in his bedroom and her home was not a bordello. The sound of his father's voice saying the words "damn" and "racialist." Martin looked up and saw the old woman staring at his father blankly. Uncomprehending. More words tumbled from his father's lips. Of human sympathy and love. Of race prejudice. Of Christian hypocrisy, Victorian prudery, and silly, self-righteous ignorance. Of liberty and the drives of human nature. Then Will tried to recruit Martin to his lecture by demanding he tell her what biology had to say about whether half-breeds even existed.*

"Martin, tell her what nonsense all this is!"

"Be quiet, Dad!"

It was his own voice. So loud he winced.

Will forgot Mrs. Macleod and looked at his son. He fell silent, trying to process the sound and content of Martin's unexpected command and the exhaustion betrayed by his posture. Mrs. Macleod took advantage of the silence to insist, once more, that Will had one week to find another place and left the room. Will shut off the radio and sat down in a chair opposite Martin with concern on his brow. Martin raised his head and looked his father in the eyes.

"Is what she said true?"

"Heavens, no. She's a mixed-up old maid. Fine example of Christian compassion, that!"

"No," Martin pressed, irritated by his father's reduction of what had taken place to grand principles. "About a woman sleeping here?"

"Rebecca spent the night last week. Hospitals won't hire her because she's too brown. Bless her heart, Marty, she was so upset, I couldn't let her go home. She was too embarrassed to see anyone. That's why you didn't know. I thought we were pretty quiet." He chuckled. "Guess we weren't."*

"Jesus Christ, Dad," Martin implored, resting his head in his hands once more.

"Hell, what are you upset for, my boy? It's me she's kicked out. I'll be fine. I'm barely here anyway."

Martin couldn't explain that after weeks of self-denial and a summer that would demand even greater discipline, he envied his father. For the first time in his life, he coveted his dad's ability to embrace a crusading philosophy and trust his instincts wherever they might lead. Most arguments that sexual desire must be corralled meant nothing to his father. He could even, when pressed, turn the matter of succumbing to desire into a question of justice, on the grounds that sex for conception only shut out all who could not have children from a life of love. Martin couldn't tell his dad, now, that he envied the embraces such a philosophy justified.*

Clearly the alcohol hadn't worked. He remembered everything Frank had said. All his explanations. In that moment he realized grief, not alcohol, had obliterated so much from his memory in the wake of his mother's death.

"It wasn't right, you know, what she said about Rebecca."

Martin looked up again. Will's brow betrayed that he was hurt, even miserable, as Mrs. Macleod's words ran through his mind again. Martin reached out and pressed his dad's hand.

"I've got to go to work," he said after a moment. "We won't be able to find you a place before I leave for the station tomorrow. You best come with me."

"Don't you worry about that, Martin. I've got a place to go. She's been asking me for months now anyway. You don't mind, do you?"

"No. Of course not."

"You sure you want to stay here?" his father added, punching one of the beaded pillows. "With such an ignorant killjoy?"

"I'll figure that out when I get home in August."

Martin stood and his dad watched him as he had to pause and close his eyes against the pounding pain in his forehead.

"Christ, Marty. Did you get caught up with the Wobblies again? Here, take my arm."

"I'll be all right." Martin tried to smile. "This one's self-inflicted."

His father was immobilized for a brief moment by that confession. Then he moved into action. "Come on then. Let's get you upstairs."

He put a glass of water in front of Martin as he sat down at their little table. Then he moved about the kitchenette making a pot of coffee and a sandwich. Even after placing it all before Martin, he waited some moments before beginning an interrogation.

"How's your head?"

Martin, having rested his head in his hands again as his dad worked, tried to nod. "All right." He looked up after a moment and tried to smile again. "Serves me right."

"Oh don't go all self-flagellant on me," Will replied, sitting down across from him. "But what gives?"

Martin stared at his dad's stack of newspapers and magazines piled high on the table. Somehow his dad could jumble them all together when they cleared the table to eat and then quickly locate things again when spreading them out.

"I think I've been rejected on eugenic grounds," he said to the chaotic pile.

Will breathed in sharply and his brow furrowed.

"Don't tell me someone has measured up your characters and found them wanting?"

"No. The other way around."

"Well sweep her off her feet, my boy! Do something!"

Martin looked up at his father.

"Oh dear, no, that won't do, will it?" his dad muttered as he saw the anguish in his son's eyes. He fell silent for a long moment, then said in a gentle voice, "Langdon-Davies has this great line where he insists the smart girl of the future will still say I ought marry you because you're intelligent and I ought not marry you because you're susceptible to tuberculosis; but I shall marry you because I love you and rather like the way you laugh. I always hoped he was right. I thought the girl of the future would know that the right thing to do, the decent thing to do, is to love who she loves, whether we know everything about how germ cells sort or not."*

"I don't care much about the girl of the future," Martin said. "I care about her. Now."

"What a sorry biologist you are then!" Will replied, but he smiled with undisguised pride. Then he paused and took a deep breath. "Surely if she loves you enough it won't be so easy to leave you. Not if she's a remotely rational creature and has any brains."

Martin rested his head in his hands again and groaned. Will suddenly laughed, because he didn't know what else to do.

"Gee whiz, Marty. I wouldn't have imagined you the hero of such a romance!"

—

By noon Martin was helping Pete check the final equipment list for the summer session at the biological station prior to loading everything up in one of the university's Fords. For much of the day the pain in his head served as a quite objective reminder of why he had given into medicating his cares away only once before. His headache was gone when they all stepped onto the boat in Anacortes the next morning to make the crossing to San Juan Island, but he didn't feel quite himself again until he felt the gentle breezes of the sound on his forehead. He stood on the deck for a long time, eyes closed, breathing in the scent of the sea and listening to the waves against the bow of the boat.

When he went inside Trevor proudly showed him a clipping a friend had sent him from a newspaper in Kentucky. News of the impending Scopes Trial appeared on the same page as a description of Trevor's efforts to introduce Japanese oysters to Puget Sound and save the oyster industry.*

"That's how we should be fighting Bryan," he said. "By focusing on good scientific work. The kind that produces knowledge that can be put to use in the world."

But Martin was having trouble focusing on anything beyond the fact that Helen and Josiah would be arriving at the station soon, and he would have to face her every day with what Frank had told him echoing through his head. Trevor had asked him to prepare the two rooms on the south side of the dining room for them. The rooms were flooded with light from a wall of windows providing views of the harbor. Normally Trevor, Martin, and Ben used them as offices and storage, but they had moved all the books and cabinets down to the labs so Josiah would be close to the dining room and lavatories, and Helen would be close to Josiah. Trevor had recruited two laborers from town to put up curtains, and Martin spent an afternoon mopping and dusting after the workers finished to get the rooms up to Helen's standards.

Then he packed a notebook and a book by ecologist Victor Shelford called *Animal Communities in Temperate America* in his satchel, and walked to one of the ponds hidden in the woods north of the station buildings. Having set aside the eugenics textbooks, he had immersed himself in a stack of economic entomology journals, trying to work up a set of lectures for the summer course he had promised to teach for Slater after the end of the summer session. Those books had led him to the latest articles on biological control in a new journal called *Ecology*, where he encountered Shelford's calls for more careful observations of behavior, accurate species identifications, and detailed descriptions of animal and plant communities. Ecologists were trying to map, and even experiment with, animal and plant communities in order to understand the interactions between them, but they were doing so, Shelford warned, with totally inadequate information regarding the creatures' life histories and taxonomies. Helping provide that information would entail a departure from Martin's work on dead specimens at his desk, but not his training in careful observation and testing generalizations. So, sitting at the edge of the largest pond, with Shelford as his guide, he jotted down some notes on what it would take to complete a detailed ecological study of dragonfly predation.*

He knew some of his colleagues would disapprove. Older naturalists liked to quip that ecologists say things everyone already knows in a language no one can understand. But Martin had made up his mind. By answering Shelford's call for more and better observations, he could continue his constant appeals for students to "look again, look again." He couldn't build the kind of collection required to do this with species boundaries. And if he must get pulled into applied work, he wanted it to be on the broad grounds built by ecologists rather than pesticide companies.*

He would need the refuge of that pond over the next few weeks. The next morning he met Helen and Josiah at the Friday Harbor dock, a day before the students were due to arrive. Louise and little Marjorie and Barbara walked off the boat as well, which meant Martin had to make introductions and then watch as Helen knelt down and gently cajoled Barbara into coming out from behind Louise's skirt.

It was five o'clock in the evening and Martin had expected Josiah to be weary from the day's travel north so he had already arranged for the island's only cab to be on standby.

"Do you want to take the cab with them?" Martin asked Louise. "I can walk."

"Oh no," Louise said. "A walk'll wear the girls out more than a cab ride."

"Let's walk too," Josiah said. "If, that is, Helen doesn't mind me leaning on her arm for a half hour more. She must be weary of my constant queries of 'Where are we now?'"

"You can ask the professor now." Helen smiled at Martin as she spoke.

This was the first excuse Martin had for looking her in the eyes since he had driven down to Tacoma and picked up her boxes of books. He hadn't said much to her then, for fear an "Are you sure?" would escape from his lips. Now, he turned to put their bags in the waiting cab and give the man enough coins to cover the journey, all the while thinking how oppressive the next eight weeks would be if he didn't get firmer control over himself.

He lifted little Barbara, who was dragging her mother this way and that, up on his shoulders. Louise fell into step and conversation with Josiah and Helen, while Martin paced a few yards behind as Marjorie danced about him, trying to reach her sister's shoes.

When they came to the point at which the town gave way to a road that led up a hill toward woods to the east, Josiah stopped.

"It's quite warm, isn't it?" he said.

Helen helped him remove his coat. She removed her own as well and placed them both over her arm. She was wearing a pale green summer dress, a departure from the browns and grays she wore at home.

When they reached the top of the hill Louise reached up for Barbara and let the girls run forward.

"I'll just keep up with them," she said, waving to Helen as she began outpacing them. "Do let me know if there's anything you need to settle in."

Josiah paused as the fields turned into woods and stood with his head tilted upward, as though listening. He surprised Martin by confessing that he had often visited the island decades earlier, when he spent a few weeks a year on the Lummi Reservation as a missionary.

"The Indians on this island refused to go to the reservation," he explained. "Lost their federal recognition as a result, of course, and all so-called privileges and rights thereto." He paused. "I wonder how many are left."*

This revelation knocked what the color of Helen's dress did to her eyes out of Martin's mind at last.

"I didn't know you were a missionary. Or that you worked on one of the reservations."

"Ten years" was all Josiah said in reply.

"There are some families who live on Mitchell Bay," Martin offered after a moment. "Sometimes they reef fish in front of the station."

Josiah acknowledged Martin with a nod, and then he took in a deep breath, as though to catch the memories in the breeze coming off the fields, and began walking again.

Josiah had never placed his criticisms of Darwin's talk of savages or higher and lower races in terms of his own or any other individual's past. Martin had learned this was part of a pattern to his friend's thought: that as intensely as he debated any point, it tended to be in the abstract, summarizing other men's ideas and trials rather than his own. The only exception had been when he confessed his collection of ichneumons and spoke of parishioners who had lost children on his watch. But he had quickly moved the scale of conversation back to the realm of abstract ideas and grand intellectual debate, by asking Martin to read the passage from Reade's book.

When they arrived at the dining hall and Martin introduced them both to Trevor and Ben, Josiah brushed off Trevor's apology for backing out of the arrangement to buy his collection with a wave of his hand.

"I know the pressures you are all under," he said.

Before he left them that evening, Josiah turned toward where Martin sat beside him and said, "We've been reading a bunch of essays collected by the modernist minister from Chicago, Shailer Mathews. *The Contributions of Science to Religion.* Pretty sure it was all pulled together to counter Bryan's crusade. I've got a long list of questions to throw at you. I'm eager to know what you think of the biologists' contributions."

"I saw it reviewed in *Science* as the best thing of its kind," Martin offered, and Josiah laughed.

"Ah, but that tells us very little about what Professor Sullivan thinks. Helen do give him our copy so he can read it and tell us what he thinks."

Josiah's use of the word "us" hurt. He had sometimes spoken like this when they were reading *The Descent of Man,* when Helen had betrayed no signal she was, in fact, listening to or cared what he thought. Martin managed a casual "Oh dear. Do I have to?"

Helen extricated herself from a game of hide-and-seek with the girls long enough to retrieve the book from Josiah's bedroom.

"I don't think you'll like it," she said as she handed it to him.

Before Martin could answer, little Marjorie called out from her hiding place behind one of the fireside couches, impatient for Helen to return to their game.

Later, after everyone had gone to bed, Martin lit the gas lantern in his tent, took it to the little table alongside one of two cots, lay down, and opened the book. He read Mathews's argument that evolution proves God is good. Because the progressive development of humanity's ideals, purposes, and individual personality over time could be tracked in the historical record. Because God's goodness could be discerned in the development over endless ages of the Golden Rule. And because man had evolved from a mechanism-bound animal to a being possessed of good will, who just might be able to obey it. Martin wondered what Helen had thought as she read the lessons Reverend Mathews drew from nature and history. Did this, too, remind her of her medieval bestiaries?*

He turned back to the table of contents, looking for contributions by biologists. The opening chapter was by William Emerson Ritter, a biologist from California, and entitled "The Method of Science." The goal of science was quite simple, Ritter explained: to see, hear, touch, and smell well enough to remove the last trace of doubt as to whether one's conclusions about the natural world match reality. It was a high standard, demanding more and better observations. And it was a value, Ritter insisted, that was just as important in life as a whole. Such a value drove a man to correct illusions. To verify first impressions. To test guesses or suppositions. And to closely examine the grounds of his beliefs before acting on them.

Then Martin turned to the other contribution by a biologist: a chapter on eugenics by Charles Davenport. The contrast was stark. Had Mathews not seen the differences, Martin wondered, as he gathered up these essays and bound them together to demonstrate the contributions of science to religion? Davenport pronounced that biology proved that inequality within human societies was due to heredity rather than differences in opportunity. Indeed, the inborn, ineffaceable inequality of human beings, he wrote, is the one great, fundamental, biological fact of society. Ignorance of this fact caused the waste of untold millions in trying to make people alike and resulted in blindness toward the only method by which most social evils would be cured: proper mate selection. He denounced institutions for the feebleminded, where hundreds of "children" (Davenport put the word in quotes) were bedridden for life. Physicians who "should have known their duty better" had prevented the congenitally deformed and senseless from dying because of some perverted instinct that valued life rather than death. When biology clearly proved that death was nature's primary means of purifying the race.

Josiah had said they had not read all the chapters. Given how he had avoided all mention of Darwin's eugenic passages, Martin wondered whether Josiah and Helen had an understanding about such things and Helen had skipped Davenport's contribution. Or had she read these eugenical arguments that death improved the race and was therefore beneficent?

Unable to sleep, Martin walked to the laboratory down by the shore before dawn. A few days before, he had helped Ben with some experiments on how environmental changes effected the behavior of little shore fish called blennies, and had noticed a stack of textbooks on genetics piling up in a corner. There was a small volume on top by Herbert Jennings, a biologist he knew, thanks to his father, as having testified against the Immigration Act of 1924. He took it in hand and read Jennings's argument that some of his colleagues were wrenching unwarranted conclusions out of careful laboratory work. That the realization of what we cannot do is as necessary for correct guidance as a realization of what we can do. That every pair of human parents contained thousands of pairs of the packets of chemicals on which development depended, which could combine in millions of different combinations. And that no scientist knew enough to predict the outcome of any particular mating.

Yet even Jennings concluded that eugenics might be useful to the effort to improve the human condition and prevent suffering, if only scientists knew more. Martin knew that even so staunch a critic of eugenists' warnings of "race suicide" as Franz Boas wrote that the proper field of eugenics was to prevent unions that would lead to the birth of disease-stricken progeny. And these concessions meant Helen's decision, if Frank was right, was still theoretically warranted.*

A "tsk tsk" was whispered at his shoulder and he looked up to see

Phoebe smiling at him. She was supposed to be nearly three thousand miles away in Morgan's lab. Astonished, Martin transferred the book to one hand to greet her with a handshake. She laughed at that idea and gave him a hug.

"I thought you were in New York," he said when she let him go.

"I was five days ago," she answered. "But I couldn't miss a summer here. I got in late last night. Walked here in the dark. What in the world are you doing reading about genetics?"

"Just trying to keep up."

He pulled up two chairs so they could sit down and asked her questions about her time at the fly lab. For a time he was distracted enough by her newfound enthusiasm for experimental work on chromosomes, but eventually their conversation from the previous summer and the memory of Helen's boxes of books inspired him to ask whether the lab men treated her all right.

She looked at him with a furrow to her brow that indicated she knew quite well this was an uncharacteristic question.

"You mean despite the fact I'm a woman?"

Martin smiled slightly and nodded. "Yes. Despite that."

She paused for a moment before replying.

"Morgan's no trouble on that count. He actually encouraged me to do independent research, rather than count flies for one of the boys. I'm not sure that indicates much progress. I've heard too many stories about what happens when his students go out into the world. Do you know Herman Muller? University of Texas hired his wife Jessie when he got a job there. Fired her when she had a baby. Administration decided a mother can't give full attention to classroom duties and still raise a child. She's having a hard time. Herman's as mad as a hatter about it."*

And then, because Phoebe was a good observer, she detected something in Martin's manner and question she had never seen before.

"You all right, Martin?"

Normally he would have deflected her concern. But Phoebe had just come from one of the best genetics labs in the country.

"Do they ever worry about how this work is used outside of their laboratories?"

"How do you mean?"

"Davenport and all his talk about a new eugenics conscience. Ladies choosing a man based on his IQ rather than his laugh."

"Oh yes, I've read my Davenport," Phoebe replied. She picked up one of the journals on the table, idly thumbing through it. "Muller insists one can't have real eugenics under a capitalist system, though. Level everything, now, absolutely level it, and one might just be able to say what's heredity and what's not. Then, somehow, you've got to figure out who to trust to determine the grounds on which men and women should be sorted. He thinks as things are now we'd end up with a population composed of a maximum number of Billy Sundays, Valentinos, Jack Dempseys, Babe Ruths, even Al Capones, since that seems to be what the people want. He's had a book on the subject in the works for a while now."*

"But what about Morgan?" Martin pressed.

"Morgan won't touch any of this," she replied, shaking her head. "He resigned from the American Breeder's Association because members made such reckless statements about human breeding. Says all this talk about feeblemindedness being inherited as a Mendelian trait is nonsense. That surely social conditions must be considered. But Morgan won't criticize them in public. He avoids discussions of eugenics and religion on the grounds he can't speak on such subjects with authority. That's his excuse anyway."*

She glanced up at him and smiled, as though to say "What's yours?" once again.

"I suspect they regret having scurried back to their labs," she continued. "Morgan was pretty upset about how claims about the genetic capacity of southeastern Europeans and Asians was used to justify the immigration act that passed last year. But they left the discussion of human heredity to the likes of Harry Laughlin and Madison Grant. What did they expect would happen?" She paused for a moment. "I'm often asked how long I'm going to work at Morgan's lab. You know what's behind the question? When am I going to start having babies, because apparently I'm the right type for it. College educated and all that. Eugenists talk a lot about the fall of civilization if the good of the individual is placed before the good of the race. But maybe, if the welfare of the species is inimical to the welfare of the individual, then the species had far better die out."*

She turned her head slightly, as though to observe a different angle of his face.

"You sure you're okay, professor? Really isn't like you to worry about any of this stuff."

"No. It isn't. But maybe it should be."

Phoebe's brow furrowed, as though she might argue with him, but then she just laughed. "Maybe."

After Phoebe left, Martin stared at the stack of books before him. When he told his dad he'd been spurned on eugenic grounds, Will had expressed surprise that he was the hero of a romance. But Martin did not feel much like a hero. Or as though he was in a romance. To be in a drama surely required some action. But all his training drove him to stay his hand if he suspected a wish to see things a certain way was driving his conclusions. That was how he maintained control over how he interpreted nature. It was how he made decisions regarding species boundaries that had never been questioned by another taxonomist. It was how he maintained control over his moments. But his interests had become so enwrapped in his assessment of geneticists' arguments that he distrusted his ability to think impartially, much less speak in the interest of truth rather than persuasion.

He heard the sound of conversation and footsteps and realized the students had arrived, having been guided in their travels north by Pete and a few other graduate students. Trevor had confessed to Martin that he hoped they could all focus on nonhuman animals, with no application to anything for the entire session. But as Pete, Trevor, and Martin helped students find their tents and get settled, Ben pulled up to the dining hall in the island's cab, having snuck into town, and proceeded to unload a new radio. The students cheered. No doubt they had visions of dancing to the latest popular songs rather than a few oldies plus the Charleston played on the station's old, out-of-tune piano, but Ben had other ideas. The rumors, it turned out, were true: WGN in Chicago would be broadcasting the trial of John Scopes to the nation, a first for American radio. Trevor grumbled something incoherent and stalked off to the tide pools after watching Ben and two students lug the radio into the dining hall.*

When Martin walked to the dining hall that evening, the sound of "Yes Sir, That's My Baby" drifted from the radio. He slowed his pace when he could see through the windows. The building was lit up brightly and the students were dancing. He caught a glimpse of the couches by the fireside and saw Helen and Josiah. If he asked her to dance, he could touch her, put his hand on her waist, and hold her fingers in his own. The thought very quickly turned into a resolution to do no such thing.

"Professor Sullivan" he heard behind him. He turned to find Pete and Phoebe smiling broadly.

Blood rushed to his face, until he determined their smiles had nothing to do with the content of his own thoughts. They were arm in arm.

"We've got to tell someone and you're the first friend we've come across," Pete said. "Guess what!"

Martin laughed. He had never seen his even-keeled young friend so animated.

"We're walking down the middle aisle, that's what."

"Good grief!" Martin muttered in surprise, and quickly extended his hand. "That's wonderful, Pete. Congratulations."

"It's all quite irrational," Phoebe smiled. "But I like the way he laughs."

Chapter 18

Amid all their hours discussing Darwin and working through Josiah's collection, Martin and Josiah had never been in the field together. Now, Martin was astonished by his friend's ability to identify living organisms. He knew every bird and frog call, and named plants and insects from brief descriptions, often sharing a memory about the creature's life history from when he could observe them by sight. He gave no sermons, even on Sunday. As he walked the woods and shore with the students for hours, guided by Martin or Helen, he made no mention of God, the Creator, or religion, unless a student asked. And when someone did, the answer was often unexpected. When one student asked whether he believed the Bible an infallible, trustworthy guide to salvation, he answered without hesitation.

"I believe the Bible is trustworthy evidence of the human experience of God."*

"That's all?" the student pressed.

"That's all," Josiah answered. "I used to think it wasn't very much either."

Josiah's gentle manner, his deep knowledge of natural history, and his reverence for both knowledge and the natural world, given the students knew he was a minister, proved a helpful antidote to the fight soon emanating from Ben's radio. The station was a week into the summer session when the trial started on a Friday morning. Given how many students were eager to listen, whether to hear the novelty of a trial broadcast live, or because they cared about the grand issues at stake, Trevor had agreed to postpone some morning field classes as long as the circus didn't go on too long. But the opening of the court proceedings proved a disappointment to those expecting a good row over the radio waves. After a long prayer by a Dayton minister who appealed for the presence of the Holy Spirit to be with the jury, the accused, and all the attorneys, the judge, a man named Raulston, read the anti-evolution act passed earlier that year:*

> Be it enacted by the general assembly of the state of Tennessee that it shall be unlawful for any teacher in any of the universities, normals and all other public schools of the state, which are supported in whole or in part by the public school funds of the state, to teach any theory that denies the story of the Divine creation of man as taught in the Bible, and to teach instead that man has descended from a lower order of animals.

Then, speaking in a firm, clear voice, he read the first two chapters of Genesis into the record.

"Argh!" escaped from Ben several times during the opening statements. The defense explained that they would defend John Scopes by bringing forward expert testimony from both scientists and theologians to prove the grave error encompassed in the law's assumption that evolution could not be reconciled with Christian faith in the divine creation of man. The prosecution replied they would be asking that such testimony be barred as irrelevant to the question of whether Scopes broke the law. But for most of the morning, the court debated tedious intricacies of jury selection. By the time potential jurors were being interviewed that afternoon, most of the students had drifted away.

Ben, Josiah, Martin, and Pete stayed. Indeed, Martin surprised Trevor over the course of the trial by sitting by the radio just as much as Ben. That first morning, Martin was captivated by the repetitive pattern of questions asked of prospective jurors. In his first task for the defense, Clarence Darrow asked each man, whether farmer, merchant, minister, or teacher, whether "your mind is in such shape that you could be perfectly fair, dispensing with any bias, prejudice, or passion, and try the case with an open mind, based solely on the law and evidence."

"What a farce," Ben grumbled. "That's a tall order for a jury made up of a bunch of southern yokels."

Josiah laughed. "I'd say it's a great deal to ask of any man, no matter where he comes from."

Martin had worried about Josiah having to put up with Ben's various speeches but they got along well enough. At times Ben even seemed a bit in awe of Josiah's tolerant manner of responding to his occasional outbursts on how Christianity had interfered with all progress. But sometimes the topic of conversation tested Martin's patience, whether Ben seemed in awe or not.

"I bet Bryan's chomping at the bit for his turn at the bullhorn now," Ben said now. "Eh, Reverend Gray?"

"Are you rooting for him?" Josiah said with a smile. "Nice to have a worthy enemy?"

"Good God, no. We'll never be able to convince men and women to stop worrying about heaven so much and improve conditions on earth with medieval priests like him around."

"But you're enjoying the circus."

"That depends how it all ends," Ben replied. "All this rubbish about science describing nature and the theologians getting everything else. It's damned irritating."

"Surely you didn't expect us to just roll over and leave the entire field without a fight?" Josiah said with mock astonishment in his voice.

The weekend was full of baseball games, trips on *The Medea* for sea cucumber lunches, hikes on Orcas Island, and evening dances. It was all a fine respite from controversies in a courtroom on the other side of the country, but at dinner Ben read the journalist H. L. Mencken's dispatches from Dayton for Baltimore's *Evening Sun*. Ben relished Mencken's acerbic division of the human race into two mutually antagonistic classes, "almost two genera," the intellectual minority that ensured progress and enlightenment, and the mass of ignorant vermin who find new ideas painful. Trevor put a stop to any more exposure to Mencken's secular sermons with a curt pronouncement: "It's bad enough we've got a radio in the dining hall. If anyone wants to read the newspapers they can row into town and sit on a street corner."*

But Trevor couldn't turn the radio off. On Monday morning a lawyer for the defense named John Neal argued that the indictment of John Scopes be quashed on the grounds the Butler Act violated the state constitution's requirement that education be cherished, that no preference be given to any religious establishment or mode of worship, and that the Genesis story of creation was interpreted a million different ways by a million different men. The assistant attorney general Ben McKenzie replied for the prosecution: "Under the laws of the land, the constitution of Tennessee, no particular religion can be taught in the schools. We cannot teach any religion in the schools, therefore you cannot teach any evolution, or any doctrine, that conflicts with the Bible."

When Darrow took the floor a crowd quickly formed around the radio as his deep voice emanated from the corner of the dining room. Martin knew of Darrow's closing statement against capital punishment in the Leopold and Loeb murder case. Everyone did. But no one had actually heard him deliver one of his famous speeches.

It was mesmerizing oratory. Here is a country, Darrow said, made up of Englishmen, Irishmen, Scots, Germans, Europeans, Asiatics, Africans, men of every sort and men of every creed and men of every scientific belief. Who, he asked, is willing to sort these men out and declare: "I shall measure you; I know you are a fool, or worse; I know and I have read a creed telling what I know and I will make people go to heaven even if they don't want to go with me, I will make them do it." Darrow did not defend evolution, or human beings' common ancestry with apes. He defended the right of a man to think differently, and argued that the Tennessee law undermined that right by insisting on a particular interpretation of the Bible. Darrow's voice over the crackling static of the radio was stern and grave:

Today it is the public school teachers, tomorrow the private. The next day the preachers and the lecturers, the magazines, the books, the newspapers.

After while, your honor, it is the setting of man against man and creed against creed until with flying banners and beating drums we are marching backward to the glorious ages of the sixteenth century when bigots lighted fagots to burn the men who dared to bring any intelligence and enlightenment and culture to the human mind.

The audience in both Tennessee and Friday Harbor broke out into enthusiastic applause. Martin remained very still and looked at Josiah. Darrow had said nothing of Christianity as the enemy. He had even, in reading the clause from the Tennessee Constitution, insisted that "all men have a natural and indefeasible right to worship Almighty God according to the dictates of their own conscience," paused, and added, "Including the modernist, who dares to be intelligent." Darrow's targets were men, ostensibly from any side, who would try to control human thought. But given weeks of the newspapermen playing the trial up as a duel between science and religion, it was hard not to envision a battlefield with trumpets and truth on only one side, as Darrow came to his resounding conclusion. Everyone, after all, thought they knew who had lit those sixteenth-century torches.

Most of the students could talk of little else but Darrow's speech when Martin finally got them out into the field that afternoon. They even pressed him for an opinion, but he parried each interrogation with a query regarding some nonhuman creature in their path.

In the ensuing days, after a protest from the defense against the judge's practice of beginning each day with an orthodox prayer, visiting modernist ministers filed a petition that the court also allow prayers from men "who believed God's divinity could be found in the wonders of the world and in the book of nature," even in the process of evolution. As a result the Massachusetts accent of Rev. Charles Francis Potter, a well-known Unitarian minister, filled the room on the fourth morning of the trial:

Oh, Thou to Whom all pray and for Whom are many names, lift up our hearts this morning that we may seek Thy truth. May we in all things uphold the ends of justice and seek that those things may be done which will most redound in honor to Thy glory and to the progress of mankind toward Thy truth. Amen.

"Does that count?" one of the students quipped. Martin saw Josiah smile somewhat absently in reply.

The prosecution challenged the defense's plan to bring forward expert witnesses and the defense argued that it needed those experts to demonstrate that science and religion embrace two distinct fields of thought and learning: that the Bible provided rules of conduct, while science studied the natural world, and thus no real conflict could exist between evolution and Christianity. It was an ambitious argument, given Bryan sat across the aisle with all his evidence that not every biologist maintained such a boundary, and Darrow sat opposite, with all his arguments that evolution told a man he must abandon Christianity to truly understand himself and his fellows.

After entering both the Bible and the biology textbook Scopes had used into the record as evidence, the prosecution brought an adolescent named Howard Morgan to the stand and asked him how Scopes had classified man in his biology class.

"Well," came the twang of a boyish voice over the radio, "the book and he both classified man along with cats and dogs, cows, horses, monkeys, lions, horses and all that."

The students around the radio laughed when Howard admitted he had no idea why mammals were mammals and wasn't even sure whether whales counted. Darrow followed that exchange with the query, "And did he tell you anything else that was wicked? Did it hurt you any?"

A stern voice broke in demanding that Darrow ask the boy's mother.

After the radio was switched off, Josiah retired for his afternoon nap, and the students boarded *The Medea* for a dredging field trip with Trevor. Martin walked alone to the pond where he was designing an ecological survey. But he did not take out his notebook. Instead he just sat and watched the dragonflies for an hour. Dozens of individuals flew about, jockeying for the best grass stems. As a young man he had become enamored of this world in which human concerns meant nothing. The dragonflies went about their business unfettered by anxieties regarding whether their moves to the right or left were wrong or right, or why they had done so in the first place. Without the benefit of humanity's caution-bound "new brain," as Trevor called it, they need not distrust their desires, and thus were completely unplagued by self-doubt. This was one reason why scientists thought their behavior might be understood. Jacques Loeb and other biologists tried to extend to human beings the assumption that animals could be understood as law-bound machines. They believed humans would understand themselves better, then, and eventually, perhaps, even be better. But Martin felt very keenly that nothing in his understanding of the little creatures around him told him what to do. If anything, the mystery of their ways only made his self-doubt worse.

He walked a trail that led out of the woods, high above the shore, back to the labs. He had not expected to see Helen or Josiah again until dinner. He thought Helen would stay near Josiah to be there when he awoke. But as he came out of the woods onto the rocky overlook where the students used to gather to watch the sunset, he saw her sitting on a moss-covered rock, looking over the San Juan Channel. He halted, tempted to backtrack and return via one of the trails through the woods. He had rarely spoken to her since their arrival, and never alone. He had his excuses: She was busy with Josiah. Or with Barbara and Marjorie, with whom she'd become good friends. Or he was listening to the radio. Or teaching a field class. But now he made himself walk forward and say hello.

It was sunny, and she had a wide-brimmed hat on. When she turned and looked up, the sun was behind him so she had to put up her hand to block its rays. He moved to stand so his shadow fell upon her face. She said hello in reply. And Martin was faced with the fact that he must either keep walking, sealing the distance between them, or say something more.

"Mind if I sit down?"

"No, of course not, Martin."

He sat down on the rocks a few feet away from her and picked up a discarded clamshell near his foot. It was not a natural place to find a clam, for this was a rocky section of shoreline high above the water. But he had a hunch how it got there. Trevor sometimes held class on this bluff, and probably carried it here as a sample while lecturing about the animals below. He held it out to her.

"See that?" he said, pointing to a small hole in the shell, near the ridgeline of the hinge. "Some snail drilled in and sucked the clam's body out. Clear case of being bored to death."

She smiled. "Oh, that's terrible."

He tossed the clam below them into the water. "One of Trevor's favorites."

She took off her hat to rub where the brim had rested on her forehead. Her temples were damp with sweat, and her long hair fell over her back from where she had wound it up unbraided beneath the hat. Martin tried to focus on the sound and scent of the waves below. But generally the waters along this shore reminded him of the night Connor died. And he associated that terrible accident with learning of Pete's troubles with his parents and reading Bryan's book. It had all influenced his state of mind when he first met Josiah and confessed the dilemma of being faced with students who had been told they must choose between God and evolution. That confession had led to friendships that now meant a great deal to him. After a few moments Martin spoke because he had wanted for weeks to show her that she need not be cut off from their conversations, just because he had made a mistake and lost his self-control.

"I've been reading Shailer Mathews's book."

"And what do you think of it?" she asked.

"Some of it's all right. I liked Bill Ritter's chapter."

"That's one out of how many?"

"I suppose I'm biased in his favor. He's always defended taxonomists from the claim real men have better things to do than describe and classify insects." She was looking at him as he spoke, so he smiled. "So you can see why I'm sympathetic."

She smiled too.

"You were right," he continued. "I didn't like the other ones. One of the zoologists from the University of Chicago dismisses creationists as ignorant, and then insists in the same breath that only trained biologists have access to the proofs of evolution. Which means, I assume, the layman is supposed to take our conclusions on faith. By the authority vested in science. A bit ironic. Especially when Mathews warns that scientists, too, can fall prey to prejudice." He paused. "I did like Mathews's reminder that in science and religion alike we are dealing, not with abstract truths, but with folks. With scientists and religious people."

She leaned forward to adjust the buckle on her shoe. "And the theology?"

"It's all beyond me. All Mathews's talk of the struggle toward freedom and self-direction as a better demonstration of God's goodness. And that a world governed by immutable laws like evolution and men able to discover and adjust to those laws is much more wonderful than a world of miracles."

Helen nodded slightly. "It is all very different from Bryan's belief that a change of heart is all that is required to reach a better earth and heaven in the hereafter."

"Both visions sound quite nice, in their own way."

"Which should have no bearing on its truth," Helen said. "You are quite religious in your application of that rule. You'd have an easier time if you abandoned it."

"Sometimes I wonder whether I'm just avoiding things a man with more courage faces head on, though he risks getting everything completely wrong. Whether he's Darrow or Bryan."

"That's your father speaking," she said. "Not you." And then she added, looking out toward the sea, "I'm surprised you sit by the radio each morning. You can't enjoy all the drama."

Martin couldn't reply that he was taking refuge in that conflict to avoid the one in his own heart, so he said instead, "No. I don't." Since she knew of his father's ways, he rambled a bit about his dad so the silence would not descend upon them again. "But he'll want to know what I think. And be very disappointed with me if I don't have at least a vague idea of what's been going on. Though I'm not sure when I'll actually see him again. Our landlady kicked him out last week."

She turned and looked at him again. "What happened? Is it serious?"

"She certainly thought it was," he replied. "I thought the police must have showed up. That he'd gotten mixed up with the Wobblies. But he'd brought his girlfriend home. Turns out the landlady is a better observer than I am. I had no idea." Helen waited, watching him, so he kept talking. "I'm not sure what upset her most, the fact he brought a woman home, or that the woman in question was part Indian. Dad called her a hypocrite because she goes to church every Sunday. And an ignorant racialist. And a prudish killjoy. I don't remember all of them. He wanted me to give her a lecture on how Darwin showed that from a scientific standpoint she was being a fool."

"And you refused," Helen guessed.

"Because I was paralyzed by a hangover in my effort to forget about you for a few hours" ran through his mind, but instead he tried to explain the vague reasoning that had kept him from obeying his dad's command. "I don't know what he expected of the poor woman. That she listen to my tales of geographical variation in beetles and abandon her belief that Indians are lower on the ladder of progress? Maybe I should have at least tried. But prejudice isn't generally something one can reason with. I'm not sure more science is needed in such cases. A change of heart, maybe." He shrugged. "She's been told another tale all her life, and I doubt five minutes with a biology professor is a strong enough antidote to cure anyone of such things."

She nodded. "Especially when a great deal is riding on the prejudice. It certainly is around here." And then, after a brief pause, she asked whether he knew why Josiah had left the Presbyterian church.

"I never knew for certain whether he had," Martin answered. "In a way, it didn't seem to matter."

"Yes. That's true," she said with a smile. "You would see that." Then she told him of how Josiah had challenged the mission leaders' ways of thinking. And their rules. "The church elders couldn't understand him. They loved stories of proud chiefs conquered by Christianity. But Josiah would ask his terrible questions in his quiet way. Was the great chief bending his knee before Christ, or before the white men who had broken treaties and stolen his land? People got pretty nervous about where his loyalties lay. They couldn't understand how his doubts that the missions were doing good might be godly rather than heretical. He got tired of arguing, and left the church."*

Martin's brow furrowed. This brief history was, no doubt, why Trevor said the young anthropologists thought so highly of him. She guessed the question running through his mind: Why, amid all their debates about Darwin's use of the words "savage" and "civilized," he had never mentioned this fight?

"I only know the stories because Frank told me. Josiah doesn't talk about it. Some of the missionaries came around to his way of thinking after the war. They had to. If they paid any attention to what had come of all their promises of the virtues of civilization and progress. But by that point he couldn't go back. I think he feels guilty for having stayed so long."*

"I appreciate your telling me," Martin said. "It makes sense of a great deal. In what he notices. And what he cares about."

They watched the waves without speaking for some time, and for the first time in weeks Martin thought they might recover something of their friendship. That maybe he could, in fact, do this. He focused again on the gentle, steady sound of the waves on the rocks, and thought of how those rhythmic waves, each different and yet all the same, had rolled upon this shore every few seconds for thousands of years since the last glacier receded, moving billions of particles of organic matter back and forth, back and forth. From dead Nereis worms to the remnants of whales. Countless individuals, their chemical bits combined to produce movement, desire, and action, broken into apparent chaos, yet grasped by new life to be formed into something else with new impulses. Even, after millions and millions of years, the impulses of love and sympathy. But then the attempt to focus on the grandeur of that history, and the insignificance of the present, reminded him of how each of Helen's moments was a moment deprived of a beloved son, whose death physicians had passed through the lens of nature's ways and the struggle for existence, and that she had been blamed. The silence in that moment suddenly became almost physically painful. Was this what Frank had meant, he wondered? When he said some eighteenth-century physician had destroyed belief in the soul and immortality and proved the purely material nature of man simply because his thoughts could influence the feelings in his gut?

And Helen? What was she thinking as Martin sat so near? She was trying to remember something in Eileen Power's book, the one Frank had given to Martin to give to her. It was the tale of a nun tempted by the devil in the shape of a young man demanding to know why she tortured herself with hunger, vigils, and discomfort, and urging her to consent, return to the world, and use those delights God created for man. The poor nun had demanded to know in reply: "How would it be with me if I should die amidst those delights, which thou dost promise me?" The devil made no reply, but eventually the nun vanquished temptation, and the devil visited no more.

Only Martin wasn't the devil. And he wasn't arguing with her about anything. So she told herself to be grateful for him. Grateful that he wasn't angry with her. That he didn't feel his manhood threatened by the fact she had moved away from him. That after just one merciless rebuke, he had stopped demanding she explain herself. That he still wished to be friends. She, too, was telling herself that giving him up was a very small sacrifice, in the grand scheme of things. Assuming one even existed. But none of this was subduing the pace of her heart as he sat quiet and still, and she was relieved when he spoke.

"Do you think Josiah will be happy living in Seattle with his sister?"

"They're actually great friends," she replied. "Her routines will take some getting used to. And her chatter. But he'll have his library."

She spoke frankly, with no apparent fear that her words might be a fine opening for Martin to ask whether she could be happy without her own books. With no knowledge of the fact that at that moment he was very close to disobeying the unspoken understanding between them, that they go on just as before. For he came very close, at that moment, to confessing he knew about David. But he struggled a moment too long to decide whether to risk another bout of her explanations. What if she replied with the earnest gaze that had vanquished all reason before?

"Josiah's probably awake now," she said an instant before his lips parted to speak. She had glanced at a watch on her wrist.

He knew her wrists well enough to know she normally didn't wear one, and asked whether the watch was new.

"Eliza gave it to me so I wouldn't forget Josiah's schedule."

She twisted her hair back under her hat. Because he sat slightly behind her, he could watch her slender fingers at work, as the loose sleeves of her dress fell back against her elbows. He looked down at the lichen-covered rocks at his feet and thought of how Darwin had written that the highest stage in moral culture occurred when we recognize that we ought to control our own thoughts. The physician and reformer Havelock Ellis warned that it was a poor sort of virtue that lay in fleeing from things that tempt us. The lesson of temptation, Ellis insisted, was to learn to enjoy what we do not possess. A child has to learn to look at flowers and not pluck them, and a man has to learn to look at a woman's beauty and not desire to possess it. That, Ellis insisted, was what constituted true civilization and "naked savages" understood the fact better than so-called civilized men.*

They walked back to the dining hall along a new path that he and Ben had made before the students had arrived, so that no one need walk over the treacherous section where they'd lost Connor.

—

After dinner, as the students danced, Martin sat next to Louise, who was repairing an injury sustained by Barbara's cloth doll during an encounter with Marjorie and a pair of scissors. Helen sat across from them with Josiah, and as she waited Barbara crawled into Helen's lap, who agreed to read one of the little children's books Louise had dispersed throughout the dining room. Josiah seemed focused on listening to the music, so Martin picked up a copy of George William Hunter's *Civic Biology* that someone, trying to be funny, had placed atop a Bible on the coffee table between them. This was the book John Scopes had used to teach evolution. In doing so he had broken the law, though the defense pointed out that the state textbook commission had approved the book for use, much to the audience's amusement. As a strategy to distract himself from Helen's voice, picking up that book backfired completely. Flipping through its pages, he came upon the section on eugenics. He read Hunter's statement that when people marry there are certain things that the individual, as well as the race, should demand. This prefaced some pedigrees tracking feeblemindedness, alcoholism, and tuberculosis through the Jukes and Kallikaks families, two of biologists' favorite examples of degenerate lines. Both the evidence and the moral, Hunter concluded, speak for themselves:

> If such people were lower animals, we would probably kill them off to prevent them from spreading. Humanity will not allow this, but we do have the remedy of separating the sexes in asylums or other places and in various ways preventing intermarriage and the possibilities of perpetuating such a low and degenerate race.*

Martin was startled by Phoebe's voice at his shoulder.

"You probably should've learned all that stuff by now, professor."

He closed the book as she came around the sofa and sat next to him. For a moment he feared Phoebe might take the book in his hands as an excuse for renewing their conversation about what men like Morgan thought about eugenics. But Phoebe addressed Helen, because she did not know the rules that governed this little circle of friends. Louise had handed Barbara her repaired doll, and Barbara hopped down from Helen's lap.

"Did you find anything at the library in town?" Phoebe asked.

"No. It's quite small," Helen replied.

Josiah raised his head. "You went to the library?"

But he said it so softly that Phoebe didn't hear him.

"My apartment's a block from the New York Public Library," Phoebe continued. "They've got everything."

She had addressed Helen entirely as she spoke, so silence was not an option.

"I lived a few blocks away from that library when I was a child," Helen replied.

"I wouldn't have taken either of you for New Yorkers," Phoebe said.

Helen just smiled again and asked Phoebe if she was going back east after the session ended.

"I've got the position for another year. Then I should probably teach elementary school or something. Morgan's offered to supervise a doctoral thesis, but I'm not sure what good it would do. Sullivan would hire me. Cardiff wouldn't. And there are a whole lot more Cardiffs in the world."

"What about a woman's college?" Louise asked.

"Tend to hire men to teach sciences, though, don't they?"

Louise nodded. "Yes, that's true."

Phoebe turned to Martin. "Think I should go into home economics? Would Cardiff hire me for that? I'd be doing my part to save civilization."

"Ben might surprise you," Martin replied. "I hope he'd know better than to argue with a doctoral diploma in biology."

Phoebe laughed. "You'd like to think that."

Within a few moments both Ben and Pete came around the side of the sofa as well and joined their little circle. Ben noticed the book in Martin's hands.

"Hey Sullivan, you aren't hiding behind that book as an excuse not to dance, are you?" Martin looked up as Ben smiled and added, "Pete and I've been having an argument about you."

Martin glanced at Pete, who seemed suddenly uncomfortable.

"Giving Pete practice for married life?" Martin ventured in an attempt to deflect the bent of the conversation. "Did you know these two just got, what do you call it these days, insured?"

Both Ben and Louise let out a "Good heavens!" while Josiah leaned forward. "Let me shake your hand, Pete," he said. "May you be good helpmates to each other. Where's your future bride? You too. Let me shake your hand." Phoebe jumped up and embraced him.

"I never thought I'd get married," she said as she let Josiah go. "In the abstract it all seems too convenient an institution for keeping half the human race in obedient subjection, while corrupting the other half with too much power."

Louise laughed. "You certainly don't sound like someone who just got engaged."

"Well, there's the abstract. And then there's Pete," Phoebe replied.

Ben apparently thought this a fine means of getting the subject back on its original tack.

"All right, now. Fine attempt at changing the subject, Sullivan. Now that's all done: Pete says you must have your reasons, and I say you can't have any good ones, unless you're not much of a hoofer and can't dance. You've been here how long? Almost two years, and I'm sure you've got a whole lot of species making done, but not a love affair in sight."

Martin's brow furrowed. He was not surprised Ben talked about him like this behind his back, but it was irritating to have such direct confirmation that he did so with students, and in this company. Ben caught something in his gaze and misdiagnosed it.

"Ah, you're astonished we talk about you like this. Well, it isn't because we dislike you, so you have that for consolation."

"If you like me so much, you'd leave me alone," Martin replied. "Judge not, and ye shall not be judged."

There was a sternness to his voice that caused Phoebe and Pete to exchange glances. Martin did not know that Pete had often tried to convince Phoebe that there were some things on which Martin would speak. Ben put up his hands in mock surrender.

"Sounds like you're angling for Reverend Gray's job. You've been training him, have you, reverend?"

Josiah smiled. "He doesn't need any training."

"Perhaps not," Ben replied, leaning toward Josiah with a mock conspiratorial tone. "I've always said he might as well be a medieval monk.

They took the righteousness of self-command too far and turned inaction into a virtue. They repressed the battle with other males and the campaign to inspire a female to choose them with a million, wasted moments meant to impress God. But I suspect the men in the monasteries missed most of what makes human life worth living." He slapped a hand on his knee and turned toward Martin. "And you don't even give the girls a chance to make a choice."

Another waltz had begun, and Josiah leaned toward Helen but spoke loud enough so Ben could hear the conversation was over. "I think I can remember a few steps before they all start doing the Charleston again. Would you mind?"

As Helen guided Josiah to the dance floor, Marjorie decided she would dance too and pulled on her mother's sleeve, so Louise joined them as well.

"I suppose a minister's daughter has to be a bit silent and straight-laced," Phoebe said as they watched Louise and Helen maneuver their partners away from other dancers. "But there's something about Miss Gray I like very much. Wonder I didn't guess she grew up in New York. She's got a certain style about her. And a bit of mystery. Maybe an adventurous past full of romance . . ."

She stopped speaking midsentence. Martin felt her look at him, though he had stopped watching the dancers and stared into the fire.

"Nothing romantic about a good woman ending up a spinster," Ben replied with a grunt. He leaned forward into his lecture mode, but Phoebe spoke first.

"Hey Cardiff," she said. "Since you're not getting anywhere with your cross-examination of Professor Sullivan, you can dance with me, if Pete doesn't mind."

Ben replied with a "Sure!"

"Sorry for ribbing you, professor," Pete said after they had gone, though Martin could not tell whether a glance had been exchanged and that statement was connected to Phoebe's sudden silence.

"You didn't say anything, Pete," Martin replied.

Pete was watching the dancers.

"Have you told your folks?" Martin asked after a moment.

"Yes," Pete replied with a smile. "They didn't like the idea, and they don't like the idea of her. But they like Phoebe. And I guess that's what gets us through these muddles, isn't it?"

Martin nodded and smiled in return.

"Yes. It does." He paused for a moment, and then added, "So they're okay with you staying at the university?"

"Well, reconciled. Not sure they're okay with it."

"I am sorry about that, Pete."

Pete laughed. "Why? It isn't you who led me astray. Or Trevor. Or Ben. Threat of eternal damnation, now, that'll drive a man out of the church. If he doesn't find his way back via more enlightened versions. That and the men and women he meets who act better than he does, though unbelievers." Pete stood up. "Well, I think that's enough of that."

Martin watched as Pete crossed the room and tapped Ben on the shoulder as the song changed. Helen was leading Josiah back to the couches as the sound of Frank Munn's "Let Me Have My Dreams" drifted from the radio.

Martin was proud of the fact that he had been able to keep up the facade of impartiality when students and colleagues were near. His reputation for ambivalence could, it turns out, be quite handy, and that was some comfort. But then, as Josiah sat down next to him with a sigh of relief, Ben was suddenly standing behind Helen and asking her to dance. Martin felt as though his heart skipped a beat. Could it do that? he wondered. Had he gained a beat of his full share or lost one? The fact he succumbed so easily to this selfish physiological effect, despite all his resolutions, was infuriating. He tried to suppress the feeling by a dozen reasons why jealousy of Ben, when he had no claim on her, was completely irrational. He knew Bryan would have explained his inability to follow through with his resolutions as somehow rooted in original sin, while Darrow would have blamed the sharp pang of unease on some vestige of his animal past, for which he could not be blamed. But Martin found little comfort in either explanation.

He made himself look at Josiah rather than Ben's hand on her waist, and had almost recovered enough to hazard telling Josiah what he thought of the biologists' contributions to Shailer Mathew's book, when Josiah knocked Martin's composure, much less grand debates about science and religion, right back out of him with one question.

"That man who asked Helen to dance. Professor Cardiff. How old is he?"

Martin had to turn his attention to Helen's companion, though he had bent all his concentration on ignoring him.

"About forty-five, I think."

Josiah fell silent.

"Well," he mused finally, "this is an interesting little experiment. Only path to getting rid of our biased ideas and developing certain knowledge, you know." Josiah chuckled at his little scientific joke. "Of course, I don't need to interpret the results. When you can tell me yourself. That's the one advantage of dealing with human beings. If they know themselves well enough, and are honest. What are you feeling just now, Professor Sullivan?"

Martin did not speak.

"Come now, Martin," Josiah pressed. "Give me an answer."

Martin finally yielded because that had always been the pattern of their conversations, no matter how difficult the question or what was at stake.

"A bit sick."

When his confession was met by silence, Martin looked up and saw Josiah's sympathetic smile.

"So. The religiously cerebral professor does have a bit of the animal within him after all," Josiah said. Martin said nothing. So Josiah added with a gentle firmness, "There comes a point when one must speak, my friend."

Martin shrugged helplessly, though Josiah could not witness it. Maybe Josiah could hear, and had often heard, the movement of his shoulders against the fabric of his shirt.

"I did, Josiah. In a way. She said no." And then he spoke aloud words that had been running incessantly through his mind for weeks. "Maybe one day she will change her mind. Perhaps she will see things differently."

"Heavens, do you hear yourself, Martin?" Josiah whispered. "And still dare to claim that you are entirely material? What illusion creates this sacrifice? This 'I ought not speak,' though it breaks your own heart? Though it renders you impotent to act, and demands of you a celibacy that promises no certain reward nor recompense except a taunting 'perhaps she will change her mind'? My dear Martin, what is it that gives you the strength to do this?"

Martin released a quiet, cynical laugh.

"Madness? Plenty of biologists would claim that I'm in the throes of some mental imbalance. Who say love is nothing more than the drive to perpetuate the species, idealized by mental tricks that allow us to imagine our desires are something above and beyond nature."*

Josiah shook his head. "I don't care about its origins. But that it exists. And that it might influence her days and yours."

Martin fell silent for a long moment. He countered Josiah's unexpected interrogation with just one of the relentless, unspoken questions that had burdened his wish to understand his friends, ever since Frank's first visit.

"Josiah, have you read her work?"

"Yes, of course," he replied, as though Martin's entire acquaintance with them did not contradict the words. And then he acknowledged the incongruity of his answer. "Did Frank convince you that I don't mind she's given it up for housekeeping?"

Martin shook his head. "He didn't say that."

"Still. I know Frank blames me for how she spends her days. But I tried to convince her to return to her scholarship. I could hear in her voice that my words pained her, so I stopped. I even asked her once why she didn't join our little debates. When I knew she trusted you. She said men generally like to be let alone to debate their troubles with other men."

"But surely she knows me better than that. Now."

Josiah nodded, but all he said was "Yes."

The toll of ignorance became too much then, and Martin broke all his long-standing rules of refusing to accept second- or third-person testimony when the beliefs or stance of someone who was not present was at stake.

"Frank said the doctors blamed the loss of their little boy on her academic life and weak heredity. He said she believed them, and that's why she won't return to her work."

Josiah's reply was so quick and firm that Martin couldn't tell whether he spoke the words from conviction or because he was trying to convince himself.

"No. She knows better than to take their little explanations seriously." He paused. "The silence that set in after David's death was very hard to witness. Frank couldn't handle it. He sent her away. It was such a selfish thing to do! He couldn't forget little Davy when she was near, or concentrate on his campaigns, so he handed her over to those damned white coats."

"I think he regrets that, Josiah," Martin offered, pained by the anger in his friend's voice.

Josiah shook his head. "And is that good enough?"

"Repentance is supposed to count for something, isn't it?"

"Should it? Once the deed is done?" Josiah replied. "When Frank came to the house a few weeks ago I said some awful things, Martin. That he be a man and let her go."

"So that's why he came," Martin said it under his breath. To himself, but of course Josiah heard.

"What do you mean?"

"He showed up in my office two days before I left Seattle. He said he'd give her a divorce."

Josiah sat in silence for some time. Then he leaned close, and whispered a brief, merciless interrogation, though his tone was as gentle as ever. "And do you love her, Martin? Do you think she loves you?"

"I don't know what I think anymore, Josiah."

Josiah reached out a hand and found Martin's shoulder. In the ensuing silence Martin finally diagnosed why they had become such good friends. Kindred spirits, even, despite their profound differences in beliefs. By some trick of nature, nurture, or an unfathomable mix of the two, they both profoundly doubted their own minds on the grand questions of the day. Martin had the fact men came from apes as an excuse. Had not Darwin asked why the mind of man should be trusted, given its origin? Josiah appealed to the biblical lesson of Job, that man could not know the mind of God. As a result, each had had to abandon a great deal of what motivated their fellows. From Helen's words, Martin discerned that Josiah had lost his ability to pronounce on things his parishioners and church elders desired he speak with confidence. Doubt had cost him the confident evangelicalism required of his profession, even among modernists. While Martin was in danger of giving up the ability to say what was good for another, much beloved human being, because his own desires were involved.

Martin thought that in the face of his dilemma Josiah would say nothing more. Then he spoke, with a voice that almost trembled with conviction.

"Admitting you know nothing is the only way you are going to find out what you must believe, Martin. You must distrust yourself completely, before you can build a sturdy foundation for true conclusions. You know that. It's what makes you so good at what you do. You have an extraordinary sense of self-command that allows you to ignore what you wish to see. In your little beetles. Even the ichneumons. You have been trained to think that following your own desires is a sign of weakness. Because you might draw the circle around your little beasts wrong. But sometimes trusting our basic decency and goodness to guide us rightly is a mark of courage, not weakness. With our fellow human beings, with our own hearts, surely, we must make a leap of faith in the end. Else you might as well be a machine after all."

Chapter 19

On Thursday morning one of the defense team's expert scientists finally took the stand. A zoologist from Johns Hopkins named Maynard Metcalf testified that it would be sinful for a biology teacher not to bear testimony on the discipline's most important theory by omitting to teach evolution. Darrow had made sure, by way of initial questioning, that everyone listening knew that Metcalf had taught Bible classes at his Congregational church and Oberlin College. Metcalf argued that God's growing revelation of himself to the human soul cannot be realized without recognition of the evolutionary method he has chosen. The prosecution responded by demanding further expert testimony be barred from the proceedings as "immaterial, incompetent, and inadmissible to the case." For the judge to rule otherwise, they insisted, would equate to announcing to the world that the jury was too stupid to determine the simple fact of whether Scopes had broken the law.

"Stuff and nonsense!" Ben exclaimed. "Damn this nonsense about some grand harmony between Christianity and science! I don't understand how men can travel all the way to Dayton, with Bryan standing across from them, and blabber on about some harmony between evolution and the Bible!"*

"What would you have them do?" Josiah asked, raising his head and turning in the direction of Ben's voice. "Not one of those scientists who traveled to Dayton is an unbeliever."

"And not one accepts the miracles of the Bible!" Ben replied. "Darrow's the only man in Dayton who might take biologists' conclusions to the place of battle where it really matters for mankind. The only man with the moral courage to tell a man straight what the implications of science are for belief." He looked at Josiah. "With all due respect, Reverend Gray."

Josiah shook his head. "I don't think that Darrow's disbelief in religion is balanced by so great a confidence in scientists."

"We haven't got anything else," Ben said firmly. "And the sooner men and women realize that fact, the sooner they'll learn that the only way to save themselves is to adjust their desires to nature's ways."

"That is quite ministerial of you." Josiah smiled with a mischievous bent to his lip.

Then William Jennings Bryan took the stage to argue against the need to place the defense's expert testimony before the jury. Everyone at the station gathered around the radio when word got out that Bryan was speaking. A loud amen from the audience in Dayton punctuated Bryan's insistence that the Christian jury did not need experts to tell them how to read the Bible, or that evolution was a beautiful thing that everybody ought to believe in. The people of Tennessee knew the dangers of the doctrine and did not want it taught to their children. Men could not teach the Bible in classrooms; therefore a minority must not be allowed to come in and teach children a doctrine that declared the Bible a lie. As for whether John Scopes had broken the law, a copy of Hunter's *Civic Biology* was all a man required as evidence. Those crowded around the radio at the biological station had to imagine Bryan brandishing Hunter's diagram of the classification of animals before the judge:

And there is a little circle and man is in the circle, find him, find man. There is that book! There is the book they were teaching your children that man was a mammal and so indistinguishable among the mammals that they leave him there with thirty-four hundred and ninety-nine other mammals. Including elephants? Talk about putting Daniel in the lion's den? How dared those scientists put man in a little ring like that with lions and tigers and everything that is bad!

Any child, Bryan asserted, who compared Hunter's diagram to the Bible's family tree could discern the conflict with Genesis.

Then he pulled out his most damning weapons against the defense team's claim that evolution did not conflict with Christianity: The fact James Leuba had established through careful surveys that more than half of American scientists did not believe in God. The fact Darrow had said that because young Nathan Leopold had read Nietzsche's extension of Darwin's survival-of-the-fittest doctrine to its logical conclusions, the young man was not responsible for murder. ("I object to the interjection of that case into this!" thundered Darrow's voice.) When Bryan demanded that Darrow admit he had blamed the professors and universities for Leopold and Loeb's actions, Darrow's reply was stern: "The fellow that intended the printing press did some mischief as well as good!" Raulston overruled Darrow's objection and Bryan continued:

> If this doctrine is true, its logic eliminates every mystery in the Old Testament and the New, and eliminates everything supernatural, and that means they eliminate the virgin birth—that means that they eliminate the resurrection of the body—that means that they eliminate the doctrine of atonement and they believe that man has been rising all the time, that man never fell, that when the Savior came there was not any reason for His coming, that there was no reason why He should not go as soon as He could.

Martin looked at Pete as rousing applause came across the radio waves. He feared it might be difficult for his young friend to hear so personal a struggle broadcast into the dining room. And across the entire country.

It was some time before the applause in Dayton died down, and one of the defense lawyers could deliver his reply. It wasn't Darrow, but a Catholic lawyer named Dudley Malone. He demanded to know whether Bryan would hold mankind forever to the limited vision of men who lived thousands of years ago and believed the world was flat. Before he could avoid the unconscious movement, Martin glanced at Helen, sitting beside

Josiah. Perhaps she had been watching him for a few moments. Perhaps for just an instant. She met his gaze and smiled. It took him a moment to recapture the bent of Malone's speech.

With as much eloquence and passion as Bryan, Malone argued that the judge must not take the defense's only weapon in the case, the scientists and theologians who would testify that the Bible and evolution do not conflict, away from them. "We have just had a war with twenty million dead," he said. "Civilization need not be so proud of what the grown-ups have done. For God's sake let the children have their minds kept open— close no doors to their knowledge; shut no door from them. Make the distinction between theology and science. Let them have both. Let them both be taught. Let them both live."

Applause in the courtroom, and in front of the radio at the biological station, continued for some time after Malone finished, until the sound of a bailiff rapping for order quieted both the courtroom and the students in Friday Harbor. The voice of the judge broke the ensuing silence with questions that it soon became clear were aimed at Darrow: Genesis said nothing of the process of creation, and many scientists and evolutionists put God behind it all, and believe in immortality. Was that right? In reply to Darrow's grave yes, Judge Raulston asked what Darrow's own view was. Darrow replied that as an agnostic, he did not pretend to know anything about it himself: "I have been looking for evidence all my life and never found it."

Martin could just imagine the shrug in the old lawyer's shoulders. A member of the prosecution jumped on the chance to drive a wedge between Darrow and the defense's argument that evolution and Christianity were not in conflict.

"Great God!" he cried. "The good that a man of his ability could have done if he had aligned himself with the forces of right instead of aligning himself with that which strikes fangs at the very bosom of Christianity."

The students booed when, the following morning, Judge Raulston read his decision to exclude expert testimony from the case. He gave the defense team's theologians and scientists the weekend to prepare their statements for filing into the record for an appeal to a higher court and adjourned early.

"They're going to lose the case," Ben grumbled as he shut the radio off once more.

"I suspect the plan all along is to get a conviction and take it to a higher court so they can get the law ruled unconstitutional," Trevor replied as he stood. He had joined the crowd encircling the radio to hear the verdict on expert testimony, to obtain some idea how much longer the circus would go on for. But now he was eager to get into the field. "Besides," he added. "They won't lose the fight if those affidavits are read into the record on Monday. Not if they're transmitted across the country like this."

—

Once more the weekend proved a welcome break from the Dayton drama, with baseball games, hikes, and dances. Martin had organized his entomology class around questions posed in Shelford's ecology textbook, so some of his students devoted their weekend hours to surveying the vegetation and insect life around the ponds. They were making good headway on the survey, but on Monday all fieldwork paused again when the radio was switched on just in time to hear the prosecution object to the defense team's plan to get the expert witnesses' statements into the record by reading them aloud.

"My objection," came the attorney general Tom Stewart's voice over the radio, "is to making a Sunday school out of this."

"It may lead to intelligent thought," countered Arthur Garfield Hays for the defense. "And that can do no harm."

"The fact that it may lead to unintelligent thought may do harm," Stewart replied.

This time, Judge Raulston sided with the defense. And so, for several hours, the Friday Harbor Bugs, and radio listeners across the country, heard the affidavits of the twelve scientists and ministers who had answered the Science Service's call to come to Dayton and fight the anti-evolution law. Even Trevor stood near, listening, as the scientists in Dayton argued in excruciating detail that evolution made sense of a range of facts, from geographical distribution to wisdom teeth. The facts in this connection are utterly senseless and insulting to an intelligent Creator, stated the zoologist Winterton Curtis, if viewed as a result of special creation: "One can simply say, God did it, and not ask why. But such explanations do not satisfy modern minds. On the other hand, their explanation in terms of evolution give reasonableness and consistency to a large body of facts." All of the experts concluded it would be impossible, one even said criminal, to teach soil science, psychology, geology, anthropology, and any biological science, without evolution.

Then the defense team's ministers spoke. Each argued against the Butler Act's assumption of a conflict between evolution and the Bible. "Since God is not subject to the categories of time and space," argued Rev. Walter Whitaker, the rector of St. John's Episcopal church in Knoxville, "a thousand years being in His sight as a single day, I am unable to see that there is any incompatibility between evolution and religion. Some evolutionists are irreligious, but so are some who are not evolutionists." Then Rev. Shailer Mathews, the Chicago modernist who had compiled the book that had bewildered Martin with its heady mix of careful descriptions of the methods of science and shrill manifestos for eugenics, insisted that "the book of Genesis is not intended to teach science, but to teach the activity of God in nature and the spiritual value of man." They could not, therefore, be in conflict. The geologist Kirtley Mather agreed. Not one fact of science, he insisted, "contradicts any teachings of Jesus Christ known to me. None of them could, for his teachings deal with moral law and spiritual realities. Natural science deals with physical laws and material realities."*

Up to that point none of the scientists had taken on Bryan's insistence that evolution equated to a ruthless survival-of-the-fittest ethics that was in direct conflict with the Christian teaching that the law of life is love. But Mather did. In reacting to Bryan's accusations that biologists were to blame for much that was wrong in the world, Mather insisted that the history of the procession of life from primitive protoplasm to man clearly showed that "at times of crisis in the past it was rarely selfishness or cruelty or strength of talon and of claw that determined success or failure. Love of offspring and tender care for the young gave the weak and puny mammals of long ago the ability to triumph over much stronger and more powerful reptiles like the dinosaur. The survival of the 'fit' does not necessarily mean either the survival of the 'fittest' or the 'fightingest.' It has meant in the past, and I believe it means today and tomorrow, the survival of those who love, cooperate, and who serve others most unselfishly."*

Martin was sitting next to Josiah. "And what do you think of that, Martin?" he said quietly.

"He just broke the rule they're all defending. That science deals with physical laws and material realities. And religion with morality and what we should value and do."

"Yes. But I'm beginning to understand it, now," Josiah said quietly. "One must, as you say, consider what he's trying to do. And that Bryan has given him no choice."

Trevor demanded everyone get back into the field after the expert witnesses' statements had all been read. But just an hour later, shouts of "Darrow's called Bryan to the stand" were heard throughout the station. Martin had escaped to his research pond, so he missed the first moments of the drama. Pete explained what was going on as Martin appeared beside him: "The defense called Bryan to testify as an expert on the Bible!"

Ben sat right next to the box, leaning in, listening closely, punctuating Darrow's best lines with a slap on his knee. Martin stood against the wall as well but could see Josiah, sitting in the place students always reserved for him on the couch, listening, his head tilted slightly up. For almost two hours Darrow examined Bryan on his interpretation of Genesis. Did he believe God made a fish and that it was big enough to swallow Jonah? Did he believe in the Flood, and in Noah's Ark, and that all civilizations had been wiped out within the last four thousand years? Bryan stood firm: one miracle was just as easy to believe as another, and in any case the Bible did not make as extreme statements as evolutionists do. He would pay more attention to the whole lot when scientists stopped arguing about whether the earth was 24 or 306 million years old.

But when Darrow asked whether Bryan believed the sun went around the earth, or the earth went around the sun, Bryan conceded that the biblical authors may have used language that could be understood at the time, instead of using language that could not be understood until Darrow was born. Though laughter from Bryan's allies came across the radio, Darrow pounced on Bryan's concession that the Bible must be interpreted. He asked whether Bryan believed the earth was made in six days, and those around the radio cheered when Bryan replied that the word "day" should not be taken literally: "I think it would be just as easy," he said firmly, "for the kind of God we believe in to make the earth in six days as in six years or in six million years or in six hundred million years. I do not think it important whether we believe one or the other."

"By God, he's got him!" Ben exclaimed.

Perhaps Bryan sensed he had conceded too much, for when the other members of the prosecution demanded the interrogation be halted, Bryan thundered: "The only reason they have asked any question is for the purpose, as the question about Jonah was asked, for a chance to give this agnostic an opportunity to criticize a believer in the word of God; and I answered the question in order to shut his mouth so that he cannot go out and tell his atheistic friends that I would not answer his question!"

The judge admonished both men at points, trying to cool tempers. When court finally adjourned, the words left hanging in the air were Bryan's insistence that "the only purpose Mr. Darrow has is to slur at the Bible" and Darrow's reply that he was merely "examining you on your fool ideas that no intelligent Christian on earth believes!"

The students were electrified by the exchange, and sat around the fireside for some time reliving Darrow's best lines before Trevor decided it was a fine afternoon for dredging and ordered them all to get down to the docks and board *The Medea*. Martin, Helen, and Josiah decided to walk to his research pond. Martin was sitting at the pond's edge explaining the study he'd envisioned to Josiah, and Josiah was peppering him with questions, when Helen, who had wandered into the woods for a space, came down the path with an ichneumon wasp in a jar.

"It was investigating the leaves of a vine maple," she said as she handed it to Martin, who then described the little creature to Josiah.

"*Coleocentrus occidentalis*," Josiah whispered. "A good find, Helen."

"Shall we take it for the collection?" Martin asked.

"No, no," Josiah answered. "Leave it be. The only life history information I've seen on this one is that it was once captured on an old hemlock log. So. We've already learned something interesting."

Martin let the wasp go. Josiah was silent for a few moments.

"*Coleocentrus occidentalis*," he repeated softly. "And out of the ground the Lord God formed every beast of the field and every fowl of the air; and brought them unto Adam to see what he would call them: and whatever Adam called every living creature, that was the name thereof."

Martin smiled. Josiah had never quoted the Bible in his presence before.

"It is said that Adam knew all the names," Josiah continued. "But that when he disobeyed God in the Garden of Eden everything became obscure and the names were lost."

"Is that what we're doing?" Martin said. "Recovering the names lost at the fall of man? Doesn't inspire much confidence that we'll get things right in the end, does it? If original sin's what's in the way."

Josiah laughed. "The theory that our most recent ancestors were some kind of hairy quadruped with a tail and pointed ears doesn't give much reason for optimism either."

Martin shrugged. "Maybe the explanation doesn't matter so much as long as we recognize the problem. And avoid deciding with certainty too soon."

Josiah tilted his head up, as though listening to the breeze, and then slowly smiled.

"I want you to have my collection, Martin. You see it rightly. I don't need the money. I'm not costing Eliza anything but a few meals a day, and she doesn't mind that. You keep the specimens."

Martin didn't know what to say. They hadn't developed an official value for the specimens since Weyerhaeuser changed his mind, but Martin knew it was worth thousands of dollars.

"He's made up his mind." It was Helen's voice. She smiled when he looked up at her. "I don't think there's much use arguing."

"Well?" Josiah pressed. "Will you take it?"

"Yes. Yes of course," Martin said finally.

"The ichneumons too. Not just the lady beetles. One must attend to both." And then he added with a smile, "So you can still do a little species making after all." He had obviously paid attention to Ben's jibes for Martin's work. "In the evenings, after your official work is done of course. Maybe get a few students to help."

"I'll try. Few are choosing taxonomy these days."

"What about Pete?"

"He's going to New York so Phoebe can continue her work in Morgan's lab. Ben's got a wager going that the geneticists will convert him, too."

"I've sometimes wondered," Josiah said, "whether the fact taxonomy's the one scientific task mentioned in scripture makes the entire endeavor quite suspect to men like your Professor Cardiff."

"Ben just likes a good fight now and then," Martin replied.

Josiah laughed. "You must exasperate him to no end." Then he raised his head. "Is Helen still near?"

She was looking under some leaves a few yards away and said so.

"You danced with Professor Cardiff," Josiah said. "Did he pick a fight with you?"

Helen laughed slightly. It was not the easy laughter Martin had once been able to inspire. "No, of course not."

"No, of course not," Josiah repeated quietly.

Up to that point the fact Josiah knew of his troubles had been a relief, though they had not spoken of Helen again. But in that moment Martin realized that even Josiah might not be willing to let him sit in silence for long.

Judge Raulston couldn't remove Bryan and Darrow's heated exchange from listeners' memories, but he expunged it from the record when court convened again the next morning. The jury, finally allowed back in the room after missing most of the week's drama, found John Scopes guilty of having taught that man is descended from a lower order of animals. The *Evening Sun*, where Mencken had filed his cutting reports, offered to pay the $100 fine. The combative tone of the previous day gave way to speeches from both sides in praise of the townspeople of Dayton for their hospitality. Judge Raulston drew the entire event to a close by insisting that the man who only has a passion to find the truth is not a complete and great man: he must also have the courage to declare it in the face of all opposition. Martin couldn't tell whether Raulston meant to praise Scopes. Or Darrow. Or Bryan. Maybe he was commending all of them, though the stands they took were so different.

For five days the radio at the biological station fell silent but for after-dinner dances. The biologists got back to work and the students fell in step beside them, carefully observing whatever organism provided the lesson of the moment. Once again, Trevor told the students they must learn to organize the chaotic buckets of life drawn from the sea and tide pools, so they could eventually see that the same principles—feeding, eliminating waste, respiration, reproduction—operate in starfish, clams, and men. Thankful that the radio no longer monopolized their days, Martin spent every free moment at the ponds. Josiah often joined him, and set to work raising various variables that he would need to control in order to eventually make certain statements about the seemingly chaotic behavior of the busy dragonflies flying about them.

Then, one afternoon, their quiet reverie at the pond was interrupted by a student running down the trail and calling out: "Hey, Professor Sullivan! Did you hear the news? Old Bryan's dead! No kidding!"

Despite the irreverent form of the announcement, and how much they had ridiculed him throughout the trial, the students and professors received the news of Bryan's sudden death, just five days after the trial's end, with solemn surprise. Even Ben seemed somewhat dampened in spirits by the news, as though everything got a lot less interesting without so ardent and irritating a combatant on the other side.

"At least he stood for something," he said, raising his glass at dinner that evening, as though toasting Bryan's memory. He coughed as the swig of plain water reminded him of how successful another of the Great Commoner's crusades had been. "Ah hell, I take it back!"

Chapter 20

A few days later, Martin rowed into town under a cloudy sky and walked up to Friday Harbor's main street to the drugstore. He had a commission from Trevor to buy a newspaper. The request had surprised him, given Trevor had insisted Ben stop bringing newspapers to the station. They'd heard via radio that the *New York Times* would be printing a closing statement Bryan had prepared for the final day of the trial. The defense had outmaneuvered him by declining to give a closing statement, and the prosecution had no right, by law, to reply. But Bryan had access to the nation's newspapers even after his sudden death.

"Only fair to let Bryan have his final say," Trevor explained in response to Martin's puzzled look, "now that this nonsense will finally stop."

"You think so?" Martin replied, and Trevor chuckled slightly.

"Well, maybe not," he said.

Soon Martin stood in the drugstore, forgetting to pay, reading the old arguments once more: that biology taught a materialistic psychology and turned man into a bundle of characteristics inherited from brute ancestors, that belief in evolution would destroy free will and moral responsibility, that Darwinism replaced each individual's ability to be born again with a damning hereditary determinism and made scientific, eugenic breeding the only logical path to reform and progress.

Bryan had added two pieces of evidence to the arguments Martin had read almost a year ago at Pete's behest: Darrow's closing statement in the trial of Leopold and Loeb, and Albert Wiggam's manifesto for eugenics,

The New Decalogue. This was the book his father had brandished before Martin with a "What are your colleagues about, Marty?" Now, Bryan demanded to know whether the ministers who insisted Darwinism could be reconciled with Christianity realized that Wiggam's book was the inevitable consequence of the evolutionary hypothesis. Bryan conceded that not all biologists made such loathsome applications of evolution to social life, but twenty-one prominent "doctors" and "professors" had publicly endorsed Wiggam's eugenic doctrines.

Then, Bryan took on the classification diagram in Hunter's *Civic Biology* once more: "No circle is reserved for man alone. He is, according to the diagram, shut up in the little circle entitled 'mammals,' with 3,499 other species of mammals." Men who were so particular to distinguish between fishes and reptiles and birds put a man with an immortal soul in the same circle as wolves, hyenas, and skunks. What shall we say, Bryan demanded, of the intelligence and religion of such men? And what must children learn from such a degradation of man?

Martin glanced up as another patron passed him so closely that he had to step aside. He happened to catch the eye of the shopkeeper. Though he had been coming to the island for two summers now, he did not recognize this young woman and she did not know him. She was clearly annoyed he was not paying his coins. He gently folded the newspaper, speech out, handed over his two cents, and walked along the wooden sidewalk down to the dock.

A light rain had begun to fall when he was in the store. The water droplets were just beginning to mix with the occasional piles of horse manure and spots of gasoline in the dusty main street, changing the smell of the air. When he came to the dock, he noticed two discarded Indian canoes alongside the rowboat he had used to cross the harbor from the biological station to town. He hadn't noticed them when he tied up. A few inches of sea and rainwater lay in their hulls, accompanied by bits of trash tossed in by passing boaters.

In the middle of the harbor he stopped rowing in an attempt to halt the world for a moment. The rowboat gently bobbed on the waves created by a passing ferry. Lummi fishermen manning two large canoes and a reef net paused their work as the wake rolled beneath them. His dad had told him stories of Indian fishermen being shot at as they canoed between the reservations and traditional fishing grounds. Martin waved, and the men waved back. Looking toward the biological station, he could see students kneeling over tide pools along the shore two hundred yards in front of him, notebooks in hand as they tried to capture with Latin names the chaotic complexity of organic beings around them.

He dreaded the ridicule and sarcasm that would greet Bryan's parting words once his colleagues and the students passed the speech around at dinner for dissection. He dreaded what might be said about Bryan's words on eugenics with Helen in the room, and about the Leopold and Loeb trial with Josiah sitting near. And he was exhausted by the charade with Helen. He avoided picking up the oars again for some time.

He could not drift in the harbor forever. Things must keep on moving, whether he had decided what to do or not. So eventually he rowed the rest of the way to the station's dock. That evening, as a rare July rain fell on the roof of the dining hall, Trevor read Bryan's posthumous speech to the group. It was digested and regurgitated with the heavy dose of ridicule Martin had expected. Ben had recovered his full disdain for Bryan, upon hearing his parting shot at all of biology. Even Trevor could barely hide his frustration with Bryan's equation of evolution with barbarism, and his attribution of the war to "this destructive doctrine." He shook his head as he read aloud:

> No more repulsive doctrine was ever proclaimed by man. If all the biologists of the world teach this doctrine—as Mr. Darrow says they do—then may heaven defend the youth of our land from their impious babblings.

Phoebe made a point of reminding everyone that women had the vote in large part thanks to Bryan's support for women's suffrage, but Ben groaned in response. "And we've got prohibition thanks to Bryan and his chiffon-skirted allies. Look at the damage he's done to science! And without an inkling of what evolution is. This is what happens when a man who believes in fables is given space to pester the American public with his medieval views."

Martin sat as he usually did, with his arms crossed and head bent slightly down, listening. No one expected him to join in.

"It's got to be said, I don't care how fresh the corpse is," Ben was saying. Sitting back in his chair with a tone of finality, he added, "The man was an ignorant ass!"

Martin had looked up. He saw furrows on Pete's brow and a bent to Josiah's shoulders that became unbearable, and he spoke.

"To call a man ignorant is a very poor explanation of his beliefs."

Ben raised his eyebrows. "And do you have another explanation of his idiotic views, Sullivan?"

Those who knew him well, including Ben, expected a shrug and silence in reply, but this time Martin answered.

"I think if we look carefully we might find pretty rational reasons why Bryan rebelled against us. For half a century Americans were told that science and reason would lead to progress. The war knocked that notion to splinters but we still crush the public with facts about ape skulls and ridicule Bryan for his ignorance, when his main concern is that men have used evolution to justify tremendous violence and human suffering. Can you blame him for taking his stand that the only safe solution is a return to orthodox Christian faith? That science and modernism had their chance, and proved tragically weak anchors for humanity's relations with one another?"*

Ben inhaled deeply, his brow furrowed.

"But Bryan and his fundamentalist monkeys are going to take us back to the Middle Ages, and destroy all science and human learning! Textbook companies are already taking any mention of evolution out of biology books. Don't tell me you're going to sit there and defend an inquisition aimed at undermining belief in the most important theory in biology. Are you serious, Sullivan? Do you want us to surrender to this buffoon?"*

"No. Not surrender. Or even excuse. But try to understand him, at the very least. It's not ignorance that inspires his campaign. It's knowledge of what men have justified by appealing to what they think biology says a man must believe and do. Concede that much of what scientists claim must be taken on faith and surely Bryan is the freethinker here, Ben. He's the man who is willing to buck scientific authority and stand against the tide of what he is being told to believe. He's the one in that courtroom who agrees with your stance, isn't he? That the reconciliations between science and religion are nonsense? The only difference between you is he sees that fact as a tragedy, and you see it as the route to salvation."

"But the man has no idea what evolution means! He hasn't the remotest conception of what he is condemning. Or the least interest in an objective study of the evidence."

"He knows enough for his purposes," Martin replied. "Bryan isn't demonstrating ignorance when he cites Hunter's *Civic Biology* or Albert Wiggam against us. And he was right about what a man might do with *The Descent of Man*. What men have done."

"And I say good riddance to his purely emotional rebellion," Ben countered. "Bryan and his fundamentalist followers are atavisms to the days when our ancestors believed in spirits because they had no other way of explaining the wind. Reason wasn't going to convince this man, or the monkeys that clamor about him as their knight in shining armor. They're all governed by emotion, rather than science. Such people may be amiable and lovable, just as is any house dog, but they are a menace to civilization. Our only hope is that they go the way of all other low-caste races vanquished in the struggle for existence."*

Martin shot an angry look across the table.

"I don't think Bryan thought much of your idea of civilization and what it must cost. It wasn't Bryan who insisted that it's criminal to hand down feeblemindedness to posterity, as though it can be tracked and predicted as easily as wrinkled peas or red-eyed flies. Or equated criminals and profligates to parasites. It wasn't Bryan who would permit the extermination of our fellow human beings because they are lower on some ladder you created."

Ben shook his head. "You've been hanging out with those Boasian anthropologists too much, Sullivan."

"No," Martin countered, trying to stave off the trembling. "I'm applying the rules we use to understand beetles, fruit flies, and guinea pigs to how we develop knowledge about our own kind."

"But you're in danger of making man an exception to natural law! You would have us give in to the trap of sentimentalism, and what we wish to believe about ourselves!"

"I will readily give into sentimentalism if it prevents me from telling another man who he is and what he should do. Or any woman." He took a deep breath to try to calm his pulse, and then continued, though the breath had calmed nothing. "Christ, Ben, is it so difficult to understand why Bryan fights us on this? He may base his conclusions on scripture, but as soon as we move from our specimen drawers, we're no better at escaping our own prejudices and desires. Do you know how many of our textbooks categorize men as though they're ideal types created by God, and then pronounce we know and can measure a man's value? This textbook Scopes taught from claims there are five races of humanity and that Caucasians are the highest type of all. Highest type? What the hell does that mean? Yet it's taught to sixteen-year-olds and called science! When it would never pass muster if we were writing about geographical variation in beetles or snails." He paused, crossing his arms and burying his hands under his elbows. "If Bryan's fight leads to a little reform and caution on our side, surely it will do us all good."

"But he did a great deal of damage!" Ben protested.

"If he did, it's our own fault. And if we let things go on like this, no one's going to trust us when it really matters. We won't do better in future if we turn Bryan into a buffoon and diagnose him as an ignorant bigot rather than a thinking, feeling human being. Flawed, yes. But so are we."

"Hear, hear." The voice was Pete's.

Ben shook his head. "It's a strange way of looking at the man, Martin."

"No it isn't." Martin could tell that the trembling was almost at his voice, so his answer was short but firm. "At least, it shouldn't be."

Someone thought it wise to attempt a diversion by skimming the rest of the paper: News that a journalist for the *Chicago News Tribune* had been kicked out of Italy for blaming Mussolini for the death of an opposition leader. A report that the former kaiser of Germany had called the Treaty of Versailles criminal and impossible. Martin soon lost the bent of the conversation. It was late, and within a few moments Helen helped Josiah to stand. As he rose, Josiah reached out a hand to find Martin's shoulder and gripped it tightly, though Helen was already steadying him well enough.

It was near midnight before Martin made his way back to his tent in the dark and the drizzle. There had not been much point in leaving the fireside earlier. He knew the adrenaline in his blood would prevent him from sleeping. After the little group had finally dispersed, he decided he might as well lie awake in his tent, rather than in the dining hall where the fire had gone out. He was standing by the little wooden desk by the canvas door flap unbuttoning his shirt in the dark, having taken off his tie and muddy shoes, when he was startled by the sound of someone saying his name. He knew that voice, and it stilled his fingers.

"Yes. I'm here," he heard himself say. "Come in."

The canvas door opened and closed and he had the presence of mind to think something might be wrong. It was the only rational explanation of her visit at this hour.

"Is Josiah unwell?"

"No, he's fine," she answered. "Are you all right?"

He gave a small laugh. "Men somehow live through war. You'd think I could bear a few minutes of debate with Ben."

"You did fine, Martin."

She could not have come to tell him that. As he waited in the darkness for some explanation, all his resolutions dissolved, and he suddenly felt angry at the stillness with which she stood so near, as though it was the easiest thing in the world for him to bear her presence so close. He decided he would say thank you for her concern and good night. Firmly. But before he could get the words out, she spoke again.

"May I show you I would choose you, Martin? If I could?"

He said nothing because he was confused by what she might mean. Then she stepped forward and kissed him. He could smell the rain in her hair, and her sleeves were damp beneath his fingers. The ability to touch her, and the realization that she, too, had suffered, almost got the better of him. But suddenly the memory of how her previous kiss had destroyed his ability to concentrate for days thundered through his head. The realization of what he must do made him grip her more tightly for an instant. Then he stepped away from her, almost knocking over the chair at the desk.

He felt in the dark to find some matches and lit the gas lantern on the table. The light flickered upon her face as she watched him. He asked her to sit down on one of the cots. He sat on the opposite one so there was some distance between them, but leaned forward as he bent all his concentration on speaking, before he lost courage.

"And tomorrow, Helen. Am I to act as though nothing has happened? As though you mean no more to me than anyone else in the room? Because I'm not sure I can do it."

"I just wanted you to know. That's all. I want you to know."

He hesitated for a moment, because he did not trust his ability to withstand the heartfelt tone of her voice for long. It was a short span of time to decide what he believed, and what he must do.

"Frank came to see me before he left," he said. "He told me about your little boy. He told me why . . . why you don't think you can choose."

"I would be careful," she said quietly.

Martin almost choked on his response. "No, no. That's not why I stopped. I don't care about that. Not in the way you think. I just want to understand what happens tomorrow. I want you to tell me. Frank said you're afraid of having children. That the doctors convinced you it was your fault your boy died."

She said nothing, gave no defense, and his heart fell.

"You believed them," he whispered. To himself. To Josiah.

"I don't know." She stared at his shirt buttons in the lantern's light. "I didn't have a reason for what happened. There was no meaning or purpose to it. How could there be? Maybe some of their commandments seeped into the breach."

"But they're wrong."

She looked up. "Do you remember when I asked you whether your agnosticism made it difficult for you to speak with Josiah? Amid all your debates with Josiah, you never wavered in your belief that the ability to doubt is a great good. Even when we crave certainty. Even when that doubt can make it seem like nothing matters."

"Truth," Martin said. He was thrilled by the candor with which she spoke, and for a moment was back in Josiah's kitchen, trying to understand her, separate from any desires of his own. Trying to explain himself, because for the first time in his life he wished to do so. "I believe truth matters."

She smiled. "I know. I was listening, remember? It's how you seduced me into speaking, because the silence was no longer comforting when you were near. Because I could see you would understand."

"I'm trying to."

When she spoke again it was with a challenge in her voice and gaze.

"Then did Frank tell you of the madness? Did he tell you I was sent away? Did he warn you that surely the machinery could be tripped again? That it could appear in my children?"

"Helen . . ."

He spoke her name as though to commence a plea. Then he stopped, in the face of the realization that the situation was worse than Frank thought. That it was the fragility of her own mind that she feared, not some education-induced weakness in her reproductive organs. He finally forced himself to speak.

"Frank never called you mad. He blamed himself. He said that Josiah did everything he should have done."

He had lost her gaze, as she stared at the floor. That fact lent a desperation to his words.

"For God's sake, Helen. You lost your son. Grief is not madness. It is like madness. But no one can say with certainty the root of the mind's fragility when we lose someone we love. You must know any claim of hereditary insanity might be wrong! Given the circumstances, it is most likely very mistaken!"

"And does your expertise as an entomologist extend to the human mind?" she replied, mercilessly trapping him within his own rules. The fact she had listened so well meant she had a great deal of ammunition against him. "Forget you care for me, Martin. Think about the evidence, selflessly and from a distance. Then ask yourself what you do or do not know about the case."

He wanted to demand of her what knowledge she had. He wanted to eviscerate the claims of the men in white coats who had diagnosed her. He wanted to do what Frank asked: use his authority and training to cast doubt on their confident judgments and explanations. But he could not now lecture her on the limits of biology. For she was right regarding the price of the knowledge science offered, mixed as it was with such uncertainty and ignorance. She was right about his inability to extend his knowledge of the hearts and nervous systems of beetles to human beings. And all his training told him not to speak when he was too self-interested in the arguments that tempted him. His sense of imminent defeat escaped in a plea.

"And what if I love you? What if you love me?"

After a moment in which he thought he would have no answer, she finally looked up at him again. Her brow was furrowed and she gazed at him in the lamplight with the earnest appeal for understanding that had unmoored his self-control and let passion triumph once before.

"You acknowledge the doubt that grips everything you do, Martin. Yet you have convinced yourself to trust whatever it is you feel and think about me. But look at your feelings from a distance! How can your reason tell you I am a safe match? When your only basis for deciding is whatever you feel when you say that you love me. Surely this is when we cannot be trusted to act rightly?"

Martin did not know whether she spoke of rightness in some eugenical sense or merely whether he really loved her. But he rebelled against her extension of every rule that guided his scientific work to what he should do in this moment. He looked at the floor and shook his head.

"Is positive knowledge required of every action? Do you apply it to every movement? Your decision to marry Frank? Your decision to stay with Josiah, and turn down the position at Vassar? We can't live like that, Helen. What knowledge could convince you by this standard that if you love me, you can choose me safely, and that I can and should choose you?"

He looked up as he stopped speaking. She was smiling ever so slightly, as though asking him to be consistent, and answer his own question. It caused the emotion rising in his throat to escape in a plea.

"I can't do this."

She gave the slightest indication of a shrug in reply, and for the second time in his life the movement seemed cruel rather than impartial and virtuous.

"Many have had to do worse."

Martin closed his lips tightly to avoid speaking. He didn't think she was giving him up for some grand vision of the future progress of the human race. He knew she didn't believe the histories on which those visions were based. But her attempt to take refuge in the claim that his or her own individual suffering was of little moment caused his anger to resurface.

Helen crossed her arms over her waist as though to ward off his arguments. He couldn't know that it was to stem the physical pain of the grief rising within her as she gave him up. But he heard echoes of the pain in her voice as she spoke. "I'm very sorry."

Martin's anger disintegrated. The careful arguments he had imagined placing before her, about the complexity of genetics and all the premature claims of those she may have believed, disappeared with the anger. So instead he made a confession that seemed completely unrelated to anything. Except that it concerned her moments, her past, and her future.

"I can't sell the books. I resolved a dozen times to load them up and take them downtown. But I can't do it. They're sitting in my apartment." He added a vague, cynical smile. "Nicely organized."

She looked up at him. Something had altered in her eyes, though he did not understand the change at first.

"But why? You don't need them."

"I thought you might change your mind. I thought your circumstances might change. That you might want them again."

There was a great deal behind the words, but he could not tell whether she understood. So he bent all his concentration on voicing what was in his own heart. Or his chemistry. Or his soul. Josiah was right. At this moment, the origin didn't matter.

"You're right, Helen. I don't know much about the human heart. I've been trained, as you have, to test every conclusion and withhold judgment as long as possible. But maybe sometimes, even in 1925, we must simply trust in each other, and in ourselves. To act on the assumption that our desires are good and meaningful, no matter where they come from, if only to get through the moments we are given."

He realized that despite what was at stake, he was not trembling. He felt an extraordinary calmness as he spoke, though she gave him neither encouragement nor opposition.

"I don't mean to claim that there is some purpose or design in it all. I don't know about that. I know that I love you. I don't care if it's just chemistry. And I know that I want you to choose me. Actually choose me. Not wish you could. I want you to have faith in me, and in our ability to meet tomorrow with as much sympathy as we can muster. That's the only thing we can do, in the end."

He stopped. The calmness began leaving him, as she sat so close and still. When she finally spoke, the words revealed nothing of her verdict.

"You don't sound very much like a scientist, Martin."

The anger escaped, and he bent forward, pressing his hands against his forehead.

"God damn it, Helen! Then listen to me as a human being! A fallible human being for whom you are willing to risk being wrong!"

His appeal was met by silence. He lowered his hands and stared at the floor between them, trying to find the words to give in. Then she sat beside him.

"Are you certain, Martin? Are you absolutely certain?"

He sensed the adrenaline rise at the prospect of trying to explain himself again. Then he looked into her eyes. He took a deep breath and, as he let the breath go, smiled at the suggestion he might, in this moment or any other, feel an ounce of doubt with regard to her. Instead of speaking, he replied by returning her kiss and yielding to desire as his sole guide.

Epilogue

As a naturalist Martin had read a lot of manifestos about the virtues and power of science. He would read many more as the decades passed. He always smiled when he did so, after that strange evening, when his lady of choice yielded to his appeal to faith in human sympathy rather than any knowledge science might give.

He sometimes wondered whether the fact he took a stand in one realm of life influenced his decision to take a stand in another. For two years he had been content to teach the courses on entomology, elementary microscopic technique, and elementary zoology, keeping diligently to nonhuman animals. But at the end of the summer, when Trevor asked the zoology professors to hand in their teaching preferences for the fall, Martin looked at the form he was supposed to fill out. He could have written applied entomology, now that he had taught the late-summer course at the College of Puget Sound. Instead he stood up and went to the storage room to retrieve the list of specimen trays still in his possession.

It was a long list. Most of the replies to his letters to colleagues asking whether anyone had the room, funds, or time to take care of more specimens had been regretful nos. But that meant that some of the best series demonstrating geographical variation were still in his possession. He drew some of the trays out of the cabinets and spread them out on the table. Some of the best were Josiah's lady beetles. His trays were full of varied specimens from which it was impossible to pick out an ideal type or form, with varieties that had once been called species but whose boundaries had been collapsed with each additional, individual specimen. These were the species that gave taxonomists intent on drawing firm circles around species and varieties the most trouble. These were the forms that led to the impression that naturalists in museums were all species makers, because how could something exist that could not be defined with confidence? To Martin, these specimens showed the complexity of evolution and of life, as forms varied within and between populations, diverged and blended back again. And he knew how to use these insect specimens to help students distrust their preconceptions, question their classifications, and resist drawing conclusions too soon.

Taking in hand the bulletin description for the department's course on eugenics, Zoology 17, he went through the notes he had gathered after promising James Slater to at least consider teaching the College of Puget Sound's course on the subject. He decided he would begin with the insect trays. Then, and only then, would they open the textbooks on genetics and eugenics. He would demand that they apply the same humility and self-doubt required for good taxonomic work to all these ambitious visions of applying evolution to mankind. After they understood the tremendous challenges of sorting a bunch of beetles, he would ask students to assess these textbooks' confident divisions of *Homo sapiens* into distinct races. After they understood how difficult it was to define and trace the origin of a particular character, he would ask them to evaluate the claims that the history and origin of complex traits like intelligence and morality had been determined. And after he had taught them how to look again, and again, and again, he would ask that they judge eugenists' confident assumption that they could divide human beings into the "fit" and "unfit" with such certainty.*

Then, lest his idealistic, hopeful students assume that one day, with enough knowledge, and enough science, biology might be a good guide to sorting the relative value of human beings, he would not assign a final exam. To examine them on these textbooks would imply a set of answers might be given with finality. Instead he would have them read an essay on evolution and ethics by Darwin's staunch ally and a founder of modern biology Thomas Henry Huxley. He would place one statement from Huxley's essay at the top of the exam:

> Evolution may teach us how the good and the evil tendencies of man may have come about; but, in itself, it is incompetent to furnish any better reason why what we call good is preferable to what we call evil than we had before.*

Then, he would ask that they answer just one question: Why do you think Huxley made this statement? So that they all pause and think about Huxley's warning. At least once.

Martin was well aware that some of his friends, who believed in banners and speeches rather than attentive silences and constant repetition of "Perhaps you are right—keep looking," would not count any of this as useful in the attempt to figure out how to apply biology to social progress. But it was what he was good at.

With a draft syllabus on his desk, he entered Trevor's office the day before the fall schedule was due and sat down, hat in hand.

"I'd like to take the Teachers' Course in Zoology, if no one else wants it."

Trevor looked up. This was a course meant for students preparing to teach high school zoology classes, and he had a hard time staffing it.

"Really? Well, you're certainly welcome to it. Takes a bit more of the theatrical element, that course, since the students in the Teachers' Course don't all come convinced of the virtues of our work. But I think you'd do a fine job of it."

"I can't promise to be very theatrical."

"No, not what is generally considered theatrical. Still. You have a reputation for being pretty inspiring. The students say you're the most passionate naturalist we've got."

Martin confessed bewilderment that any student would perceive him as passionate about anything. Trevor laughed.

"Ah, but you are. Passionate that students must be dispassionate about that which they are passionate."

Martin's brow furrowed, as he tried to dissect that statement. He decided to take it as a compliment.

"Speaking of which," Trevor continued. "I've heard a rather disturbing rumor, Sullivan." The smile indicated they would not be rehashing Martin's friendship with a bunch of Wobblies. "Gossip, really. That you were a bit distracted this summer. By a certain lovely specimen."

Martin knew he had only himself to blame for the gossip. The day after he had finally spoken, he had to navigate a field trip to Mount Constitution on Orcas Island during which Helen's attention seemed, as always, concentrated on Josiah. In some moments he feared it had all been a dream. In others, that he'd misunderstood everything. The day of distance took its toll. When he entered the dining room at nightfall as a waltz came over the radio, he had to choose between sitting down by the fireside as usual or moving with conviction and upsetting all pretense. So he imagined himself a dragonfly unplagued by self-doubt, walked with a deliberate step to stand before her, and asked her to dance. He remembered now that Louise had been sitting beside Helen. She must have witnessed what Helen's smile and the firm grip of her hand did to his countenance as doubt evaporated, and Martin finally understood why both poets and biologists spoke of love as intoxication.

"I wouldn't say just a bit," he replied to Trevor's interrogation. Then he smiled. "But I would say lovely."

Trevor laughed.

"Well! I just lost a bet with Louise! You certainly had me fooled. Some naturalist I am." He then indicated the rumor had been quite specific. "And Reverend Gray? I assume he approves of you."

Martin nodded with unconcealed pride.

"Yes, sir. Very much."

Trevor smiled. Then his brow turned a bit more serious.

"Mind if I ask a quite personal question, Martin? Just because it will be asked, and, well, I want to be able to give the facts, rather than speculations and guesses. This man Frank Gray. I've heard a bit about him. He was in prison after the war?"

"No. He went to the Soviet Union. Voluntarily."

Martin guessed that Trevor was wondering whether Helen was still married. By law she could have obtained a divorce for abandonment five years ago, when Frank had left the country. Martin had asked her why she had not done so. "I didn't care about it then," she said simply. Martin had no sympathy for the moralistic policing of faculty behavior behind Trevor's question, but he also knew Trevor was asking because he wanted to protect Martin from those who did.

"He's agreed not to contest a divorce on grounds of abandonment."

The word stuck in his throat a bit. Abandonment wasn't an accurate description of why Frank and Helen had parted, but it was the only option, since Helen was unwilling to claim habitual drunkenness or any of the worse criteria permitted by law. Of course, the courts to whom the application must be made would have found the fact Frank was a Wobbly sufficient. That pegged him as an atheist, quite enough to demonstrate that a man could not be trusted to uphold the marriage bond or any other duty. But neither Helen nor Martin wished to give the courts that kind of satisfaction.

"Well Sullivan," Trevor said. "I've always believed a man does better work when at the end of the day he is greeted by someone who believes in him. Might be a dog. Or a child. Or a parent. Or a wife. Important thing is having someone who believes that the content of your hours is worth something. Whether it leads to the salvation of humanity or not."

Martin stared at the reprints on Trevor's desk during this little speech. He was thinking of Helen's work. This was the one topic that he had not been able to broach yet. Each time a moment arose in which he might speak, the memory of Frank's challenge that she was wrong to abandon her intellectual life always halted him. He did not want to impose his own beliefs about what she should do with her time, and was afraid that if he spoke, the trembling would return, given what was at stake. For it would involve decisions about what her hours in the future would be like. And her moments meant a great deal to him now. He still waited for some signal of what she wished to do with her days.

"Of course, he probably won't do more work," Trevor was saying. "But it will be better. Now, back to business. Glad you'll take the Teachers' Course. Didn't expect you to be so unacademic. Sam Henshaw will consider you quite corrupted, you know, when word gets back to him."

"It gets worse," Martin said, as he took up his hat and stood to go. "I'll teach the eugenics course as well, if you don't mind giving me free reign to teach it exactly as I like."

"Good Lord," Trevor said, astonished, "What will your father say?"

Martin was well aware Trevor might refuse him permission to teach the course if he replied as he did.

"He's given me half the readings."

Trevor sat back in his chair. "You've read the course description?"

"Yes, sir. 'The principles of evolution in their relation to human welfare.' Would you be all right with my interpreting that quite broadly? And in teaching the course entirely as I see fit?"

Trevor slowly smiled. "There's a crusade behind that question, isn't there?"

Martin smiled as well and gave a deliberate, self-mocking shrug. "Perhaps."

He was relieved that Trevor understood. He didn't want to be underhanded about what he was doing. And he was astonished when his father not only understood, but diagnosed the two changes in Martin's life as somehow closely connected. Upon meeting her for the first time, his father embraced Helen and exclaimed:

"You're the lass who finally inspired him to look up from his little beetles, bless you!"

Helen smiled. "No. He was always watching closely. That was half the trouble."

In the end it was Martin's dad who convinced Helen to return to her work. At least, that was how Will told the story. He scoffed at Martin's "Please let her decide, Dad" stance, and showed up at their new apartment one day with two scruffy but genial friends, struggling with a beautiful oak desk salvaged from one of the law firms downtown that was modernizing its decor.

"This isn't for you, Marty. Stand aside," he said when Martin opened the door.

Helen came into the room from the kitchenette in the back of the apartment. She and Will had become good friends. Helen hadn't had a chance to maintain her pattern of silence, even if she wanted to, with Will around. When he learned she was a historian he questioned her somewhat mercilessly about her work, keeping up his long-standing habit of refusing to ignore a woman if he suspected she had something—anything—to say. At first Martin tried to protect Helen from his dad's little interrogations by changing the subject, but his dad would have none of it, and Helen didn't seem to mind, so Martin finally gave up.

Will loved to tease Helen for her belief that the past be studied for its own sake. But Helen, it turned out, could repay him the teasing in spades, by a constant barrage of "But that story isn't true, Mr. Sullivan" when he or his enemies tried to use the past in their crusades in the present. This was a playful side of Helen that Martin had never witnessed before. Their own courtship hadn't been characterized by much fun or flirtation. It wasn't until holding her close during that first waltz at the station that he had the courage to ask her questions that had nothing to do with theology or history. "Tell me about your childhood," he had said. To start at the beginning.

Will's admiration for his daughter-in-law was sealed in the final weeks of 1925. He showed up at their apartment with an uncharacteristic malaise in his face, and it was ten minutes before Martin was able to get out of him that it was the anniversary of Rebecca's son's death.

"But why are you here?" Helen said after Will told them.

"I didn't know what to say. I think I best let her be alone."

Helen shook her head. "You foolish old man. This isn't the time to overthink things. You've got imagination and a good memory. Just be with her." She crossed the room, took his hat and coat from the coatrack, and returned to press both into his hands. "Now."

Will jumped up and obeyed, well aware of how unlike it was of Helen to command so firmly.

He had been looking for a way to express what he felt about her ordering him about on this particular matter for some time. Which meant he had to do something.

"Now," he said to a bewildered Helen as they finished placing the desk by the best window. "It's there if you need it."

Martin stood, silently furious with his dad, as Will barreled into their apartment, moving furniture. It was just his way: to decide what was right, ignore all doubt, and do precisely as he pleased. Then Martin saw Helen's face as Will insisted she sit down at the desk. She didn't say anything, but Will saw the look in her eyes as well and added a delighted, "Ah, but we aren't finished, my dear!" Two sturdy bookshelves followed, placed on either side of the desk. Martin saw as he watched her that his dad *was* right about this, and that neither would have known one way or the other had Will not plunged.

It turned out she did need the desk, for within a week of the afternoon that they placed her books in the bookcases, the chair of the History Department, Edmond Meany, sent Martin a note asking whether Dr. Sullivan would be interested in teaching a course on the history of medieval civilization.

"Why would they ask you to teach a history course?" Helen asked when Martin handed her the note. He laughed.

"No. He's asking you, Dr. Sullivan."

Martin suspected his father was behind that development too: eventually he found out that indeed Theresa McMahon had told Meany about Helen's training. She must have divulged little else of her background, for Meany was one of the most conservative men on campus.

Eventually Ben diagnosed what Martin was doing in the Zoology Department's eugenics and teachers' courses. He called Martin a parlor pink to his face, though with a tinge of amusement that indicated he wouldn't be reporting Martin to the Board of Regents, or anyone else.*

The History Department hired Helen to teach classes on the Middle Ages and the Reformation each year for nearly a decade. After she had done so for a few years, Rebecca began referring to Helen's students as her children, and wrote her a note each Mother's Day with the words: "This day is for all who share their lives and wisdom with our young people. You are such a person."

"There are many ways to be a mother," Rebecca said to Martin, when he told her how much the words meant to Helen. "And a father."

Will, sitting near, found Rebecca's point a fine opportunity to turn how his son and daughter-in-law spent their days into a political statement.

"Some of your colleagues talk about genetic death, as though that's the only yardstick. But if environment and education are good for anything, you're doing your duty well enough."

Martin laughed. "Duty, is it? Can't we just call it living and leave it at that?"

To the astonishment of the historians who taught modern history, Helen's courses became so popular with students that Professor Meany asked her to teach them again year after year. Some of the students didn't like leaving a course with more questions than answers about the past. Or the fact she had an exasperating tendency to give a quiet smile when a student told a convenient tale about the past, and ask: "Are you certain? Can you find that in the historical record?" Some students just nurtured their exasperation and ignorance. Others immersed themselves in the library stacks in the face of her gentle but firm challenges.

Then, in 1935 and the midst of the Great Depression, the university passed an anti-nepotism resolution that banned the hiring of more than one member of a household. All women married to faculty, with the exception of Theresa McMahon and Erna Gunther, were dismissed. Proponents of the new rule insisted the women were working for pin money while men were struggling to find employment. They failed to mention that the resolution had been in the works long before the depression hit.*

Helen had never seen Martin so angry as when they heard the news. He wanted to resign in protest, but she sternly prevented any heroics. Trevor, astonished at the realization he needed to act quickly to prevent Martin from doing something rash, asked him to represent the university at the International Congress of Entomology in Madrid that summer. It was a strange request in an era of retrenchment, but Martin's recent work on the ecology of dragonflies—and their potential role as biological controls of mosquitos—had become well known, and the Experiment Station was willing to pitch in on travel costs so Martin could both report on his research and update the state's entomologists on the latest European work.

"I want you to leave as soon as classes end," Trevor insisted. "Take Helen, and go see the European collections while you're there."

Martin had to laugh when he told Helen about Trevor's request. Trevor clearly wanted to get him out of town for the summer. Josiah saw the irony in the idea as well. They were sitting in Eliza's sitting room, Josiah in his usual chair, and Martin sitting opposite. Helen sat on the couch nearby, reading through Josiah's correspondence for the week in case any of his former parishioners needed his aid. A week from his ninetieth birthday, Josiah moved a bit slower and slept a bit longer than when Martin first shook his hand a decade earlier, but he could still recall the placement of paragraphs in books that had disturbed or charmed him when he could see.

"Well, Martin," Josiah said, chuckling at the idea anyone found his calm friend a threat to campus peace. "It turns out you've got a bit of your father in you after all. Only trouble is I'm not sure Madrid is the best place to cool off. Especially not this year."

"Maybe he thinks a few months in Europe will convince us we've nothing to complain about," Martin replied.

Helen looked up. She was like Martin's father in one way—she read the news every day—and often read articles aloud to Martin as he made breakfast. They both knew there were reasons not to travel to Europe. For months Spain had been on the verge of a violent fight between conservatives and liberals for control of the country. Frank had sent word he was going to Europe to see how the IWW might tighten ties with the International Working Men's Association, exiled in 1933 from Berlin, and that they might not hear from him for some time. The replies of one of Martin's most reliable British correspondents were delayed because he was so busy trying to get Jewish colleagues out of Germany. That entomologist had added a postscript to his most recent letter: "At least Madrid would be better than meeting in Berlin." Just that week Martin had received a letter from a young German entomologist expressing optimism that under Hitler's rule the museum in Berlin would be able to purchase specimens again.*

"He's wrong," Helen said. "It's going to convince us to be more vigilant."

Martin did not shrug. "Then we best go."

The End

Notes

Chapter 1

p. 3 **The anti-evolution speech** For background on William Jennings Bryan's campaign, see Edward Larson's *Summer for the Gods: The Scopes Trial and America's Continuing Debate over Science and Religion* (New York: Basic Books, 1997). Bryan died five days after the trial. His final speech was reprinted in various newspapers, including "Text of Bryan's Evolution Speech, Written for the Scopes Trial," *New York Times*, July 29, 1925.

p. 5 **Two years ago** On the Museum of Comparative Zoology, see Mary P. Winsor's *Reading the Shape of Nature: Comparative Zoology at the Agassiz Museum* (Chicago: University of Chicago Press, 1991). On Samuel Henshaw, see Robert T. Jackson, "Samuel Henshaw," *Science* 93 (1941): 342–43. For the history of natural history, museums, and taxonomy, see Paul Farber's *Finding the Order in Nature: The Naturalist Tradition from Linnaeus to E. O. Wilson* (Baltimore: Johns Hopkins University Press, 2000) and Robert Kohler's *All Creatures: Naturalists, Collectors, and Biodiversity* (Princeton: Princeton University Press, 2006).

p. 7 **I'm sorry, Sullivan** For calls for scientists to be involved in public affairs, see William Emerson Ritter's writings and W. E. Allen's "The Naturalist's Place in His Community," *Science* 50 (1919): 448–51. For historical analyses of these trends, see Philip J. Pauly, *Biologists and the Promise of American Life: From Meriwether Lewis to Alfred Kinsey* (Princeton: Princeton University Press, 2002).

p. 7 **Nowadays students** The fear that most students are interested in human problems is based on Julian Huxley's "Searching for the Elixir of Life," *Century Magazine* 103 (1922): 621–29.

p. 9 **His mother, Mary** On Anna Comstock, see her *The Comstocks of Cornell: John Henry Comstock and Anna Botsford Comstock* (Cornell, NY: Comstock Publication Associates, 1953). Her *Handbook of Nature Study* was published in 1911 and has been in print ever since.

p. 10 **In 1910 Martin** On John Henry Comstock, see Pamela M. Henson's "The Comstock Research School in Evolutionary Entomology," *Osiris* 8 (1993): 159–77.

p. 10 **He learned quite early** The comparison of choosing to study insects with choosing a wife was given by entomologist Karl Jordan in "The President's Address, 1929," *Proceedings of the Entomological Society of London* 4 (1930): 128–41. Darwin's stance on the search for truth as an adequate justification of scientific research appears in a letter to J. S. Henslow from April 1, 1848, available in *The Correspondence of Charles Darwin*, vol. 4, *1847–1850* (Cambridge: Cambridge University Press, 1985).

p. 11 **Both facts were hard to forget** For the call for all American institutions, including museums, to "stand and deliver," see F. H. Sterns, "The Place of the Museum in Our Modern Life," *Scientific Monthly* 7 (1918): 545–54.

p. 11 **He could have stayed** For L. O. Howard's campaign for economic entomology, see his *Fighting the Insects: The Story of an Entomologist* (New York: Macmillan, 1933). And "(a) On Some Presidential Addresses: (b) The War against the Insects," *Science* 54 (1921): 641–51.

p. 12 **Martin's decision** For the rise of experimental biology, see Garland Allen's *The Life Sciences in the Twentieth Century* (Cambridge: Cambridge University Press, 1978) and Jane Maienschein's *Transforming Traditions in American Biology, 1880–1915* (Baltimore: Johns Hopkins University Press, 1991).

p. 12 **Martin believed Sam's** The "beast within" theories prevalent around the time of the First World War are examined in Paul Crook's *Darwinism, War and History: The Debate over the Biology of War from the 'Origin of Species' to the First World War* (Cambridge: Cambridge University Press, 1994).

p. 15 **Martin had witnessed** For Edward Aveling's exchange with Charles Darwin, see Aveling's *The Religious Views of Charles Darwin* (London: Freethought, 1883) and the analysis in James Moore's *The Darwin Legend* (Ada, MI: Baker, 1994).

p. 15 **Then Aveling's lover** For Eleanor Marx, see Rachel Holmes's *Eleanor Marx: A Life* (New York: Bloomsbury, 2015).

p. 15 **The second conflict** The calls to "ship or shoot" radical leftists are described in Frederick Lewis Allen's *Only Yesterday: An Informal History of the Nineteen Twenties* (New York: Harper and Brothers, 1931).

p. 17 **The writer W. L. George** W. L. George's classification of women is in "Analyzes Women and Has 65 Species: No Mystery about Them and He Loves Them All, Says W. L. George," *New York Times*, January 9, 1922.

p. 18 **Martin knew what** Robert Ingersoll's use of cancer to undermine the claim that evidence of a good God may be found in nature is from his *The Gods: And Other Lectures* (New York: CP Farrell, 1889).

Chapter 2

p. 21 **Hell stirred up** On Trevor Kincaid, see Muriel Guberlet's *The Windows to His World: The Story of Trevor Kincaid* (Palo Alto, CA: Pacific Books, 1975).

p. 23 **So you're our new** On taxonomists as "species makers," see Kristin Johnson's *Ordering Life: Karl Jordan and the Naturalist Tradition* (Baltimore: Johns Hopkins University Press, 2012).

p. 23 **Except Fridays** For opposition to liberal professors on campus, see Thomas C. McClintock's "J. Allen Smith, a Pacific Northwest Progressive," *Pacific Northwest Quarterly* 53 (1962): 49–59.

p. 24 **Well can you blame 'em?** Vernon L. Parrington's statement was originally made by historian Carl Becker in *The Dial*. Parrington chose it as an epigraph for an early draft of his book *Main Currents in American Thought* (New York: Harcourt, Brace, 1927).

p. 26 **Martin had a hard time** For Madison Grant, see Jonathan Peter Spiro's *Defending the Master Race: Conservation, Eugenics, and the Legacy of Madison Grant* (Burlington: University of Vermont Press, 2009). Franz Boas's review appeared as "Inventing a Great Race" in *New Republic* 9 (1917): 305–7. For the rise of Boasian anthropology, see Elazar Barkan's *The Retreat of Scientific Racism: Changing Concepts of Race in Britain and the United States between the World Wars* (Cambridge: Cambridge University Press, 1992). For the positive review of Grant's book in *Science*, see Frederick Adams Woods's "Review: The Passing of the Great Race by Madison Grant," *Science* 48 (1919): 419–20. On Grant's effort to get Boas fired, see Jeanne D. Petit's *The Men and Women We Want: Gender, Race, and the Progressive Era Literacy Test Debate* (Rochester, NY: University Rochester Press, 2010).

p. 27 **Fine. I don't begrudge** Ben Cardiff's critique of forest reserves is based on Senator Albert Johnson's testimony during the *Indian Appropriation Bill: Hearing before Subcommittee of the Committee of Indian Affairs of the House of Representatives* (Washington, DC: Washington Government Printing Office, 1920). Ben's description of race extermination as the inevitable result of expansion is based on biologist Edwin Grant Conklin's *Heredity and Environment in the Development of Men* (Princeton: Princeton University, 1919).

p. 27 **Look. I don't like** Ben Cardiff's stance on self-reliance as the source of progress is based on the work of biologist David Starr Jordan, including

Jordan's *The Care and Culture of Men* (San Francisco, CA: Whitaker and Ray-Wiggin, 1910) and *The Heredity of Richard Roe: A Discussion of the Principles of Eugenics* (Boston, MA: American Unitarian Association, 1911).

p. 29 **For weeks** On Anna Louise Strong, see David C. Duke, "Anna Louise Strong and the Search for a Good Cause," *Pacific Northwest Quarterly* 66 (1975): 123–37. On fights over education in Seattle during the Red Scare, see Keith A. Murray, "The Charles Niederhauser Case: Patriotism in the Seattle Schools, 1919," *Pacific Northwest Quarterly* 74 (1983): 11–17. On the Seattle strike and its legacy, see Jeffrey A. Johnson's *They're All Red Out Here: Seattle Politics in the Pacific Northwest, 1895–1925* (Norman: University of Oklahoma, 2008). On bootlegging in Seattle, see Norman H. Clark, *The Dry Years: Prohibition and Social Change in Washington* (Seattle: University of Washington Press, 1965).

p. 30 **At one point** On James Harvey Robinson, and the case of a teacher being fired for teaching his book, *Mind in the Making*, see Michael Lienesch's "Abandoning Evolution: The Forgotten History of Antievolution Activism and the Transformation of American Social Science," *Isis* 103 (2012): 687–709.

p. 31 **The only other guest** On the "male only" rules at the Faculty Club, see "The Woman Institute Visitor, Then and Now," *Washington Newspaper* 5 (1919): 19–22. On Erna Gunther, see Viola E. Garfield and Pamela T. Amoss's "Erna Gunther (1896–1982)," *American Anthropologist* 86 (1984): 394–99, and Lenore Ziontz's "Erna Gunther and Social Activism: Profit and Loss for a State Museum," *Curator* 29 (1986): 307–15.

p. 32 **Most people think** On attitudes toward Native Americans in the Pacific Northwest, see Coll Thrush's *Native Seattle: Histories from the Crossing-Over Place* (Seattle: University of Washington Press, 2007) and Alexandra Harmon's *Indians in the Making: Ethnic Relations and Indian Identities around Puget Sound* (Berkeley: University of California Press, 2000).

p. 32 **Maybe men** On the campaign for reform, see Randolph C. Downes's "A Crusade for Indian Reform, 1922–1934," *Mississippi Valley Historical Review* 32 (1945): 331–54.

p. 32 **Will leaned forward** On the belief that the progress of civilization is dependent on private property, see the conclusion to chapter 10 of Darwin's *Journal of Researches into the Natural History and Geology of the Countries Visited during the Voyage of HMS "Beagle" Round the World, under the Command of Capt. Fitz Roy* (London: Ward, Lock, 1889). On the impact of this assumption on government policy, see David Wallace Adams's *Education for*

Extinction: American Indians and the Boarding School Experience, 1875–1928
(Lawrence: University Press of Kansas, 1995).

p. 32   **Yes. But we're branded**   For criticisms of members of the American Indian Defense Association as Soviets, see Jon Allan Reyhner and Jeanne M. Oyawin Eder's *American Indian Education: A History* (Norman: University of Oklahoma Press, 2004). For Clallam tribe member Johnson Williams's belief that the inevitable result of evolution was each tribe's eventual extermination, see "Black Tamanous, the Secret Society of the Clallam Indians," *Washington Historical Quarterly* 7 (1916): 296–300.

p. 33   **Some refused**   Lummi men taking on the draft registrars and postwar examples of "whites only" rules are described in Alexandra Harmon's *Indians in the Making: Ethnic Relations and Indian Identities around Puget Sound* (Berkeley: University of California Press, 2000).

p. 33   **Trevor spoke up**   Trevor Kincaid's claim that the Puget Sound tribes wasted oysters when harvesting is from Muriel Guberlet's *The Windows to His World: The Story of Trevor Kincaid* (Palo Alto, CA: Pacific Books, 1975).

p. 34   **So, on the last day**   On some Boasian anthropologists' questioning of accepted gender and marriage norms, see Lois W. Banner's *Intertwined Lives: Margaret Mead, Ruth Benedict, and Their Circle* (New York: Vintage, 2010).

Chapter 3

p. 36   **It is exactly twenty years**   The statements of Trevor Kincaid and the account of the Puget Sound Biological Station are from Muriel Guberlet's *The Windows to His World: The Story of Trevor Kincaid* (Palo Alto, CA: Pacific Books, 1975).

p. 37   **That did the trick**   For examples of naturalists drawing on ascetic ideals to describe their scientific lives, see Albert Edward Gunther's *A Century of Zoology through the Lives of Two Keepers, 1815–1914* (London: Dawsons, 1975) and Karl Jordan's "In Memory of Lord Rothschild," *Novitates Zoologicae* 41 (1938): 1–41.

p. 39   **When the Lord**   The poem is from W. F. Kirby's "An Entomologist's Jubilee," *Entomologist* 26 (1893): 233–35.

p. 41   **But Linnaeus**   On naturalists, including Darwin, and the ichneumons, see Stephen Jay Gould, "Nonmoral Nature," *Natural History* 91 (1982): 19–26. Darwin's comment on the ichneumons as evidence against a beneficent God is from a letter to Asa Gray from May 22, 1860, available in *The*

Correspondence of Charles Darwin, vol. 8, *1860* (Cambridge: Cambridge University Press, 1993).

p. 41 **Trevor found a species** The possibility of seeing the ichneumons as providential means of restoring the balance of nature is based on C. W. Johnson's "Our Insect Friends," *Bulletin of the Boston Society of Natural History* 1 (1919): 3–6.

p. 42 **Martin had long since determined** On the AAAS committee, including the ties of each of these men to the eugenics movement, see Alexander Pavuk's "The American Association for the Advancement of Science Committee on Evolution and the Scopes Trial: Race, Eugenics and Public Science in the USA," *Historical Research* 91 (2018): 137–59. On scientists who joined the effort to demonstrate harmony between Christianity and evolution, see Edward B. Davis's "Science and Religious Fundamentalism in the 1920s: Pamphlets from the Scopes Era Provide Insight into Debates about Science and Religion," *American Scientist* 93 (2005): 253–60.

p. 42 **Ben had read** See T. T. Martin's *Hell and the High Schools: Christ or Evolution, Which?* (Kansas City: Western Baptist, 1923).

p. 43 **The cause of science** Ben Cardiff's response to reconciliations between evolution and the Bible put forward by his colleagues is based on the stance of Yakima zoologist Ira D. Cardiff. See Ira D. Cardiff's "Evolution and the Bible," *Science* 62 (1925): 111.

p. 43 **That's not fair** Trevor Kincaid's (and Martin Sullivan's) emphasis on detailed empirical research for which there was some prospect of solution is emblematic of many scientists' retreat from public discussions of science and religion during this period. See, for example, the discussion of Thomas Hunt Morgan in Constance Areson Clark's *God—or Gorilla: Images of Evolution in the Jazz Age* (Baltimore: Johns Hopkins University Press, 2008).

p. 43 **Ah yes** This was a critique launched at Trevor Kincaid for his tendency to avoid political debates on campus. See Muriel Guberlet's *The Windows to His World: The Story of Trevor Kincaid* (Palo Alto, CA: Pacific Books, 1975).

p. 44 **Ben sniffed** The comparison of fact gathering to avoiding "the world, the flesh, and the devil" is based on Georges Romanes's defense of speculation in *Darwin and after Darwin, an Exposition of the Darwinian Theory and a Discussion of Post-Darwinian Questions* (Chicago: Open Court, 1892).

p. 44 **Those facts aren't** For a discussion of concerns that a strong focus on amassing facts might prevent action, see the biologist Ross G. Harrison's "Science and Practice of Science," *Science* 40 (1914): 571–81.

p. 45   **Oh sure** For Trevor Kincaid's experience with an anti-evolutionist student, see Muriel Guberlet's *The Windows to His World: The Story of Trevor Kincaid* (Palo Alto, CA: Pacific Books, 1975).

p. 45   **How the hell** Various writers urged that scientific training would make better citizens, including Karl Pearson, in *The Grammar of Science*, 3rd ed. (London: Adam and Charles Black, 1911), and the historian James Harvey Robinson, in *Mind in the Making: The Relation of Intelligence to Social Reform* (New York: Harper & Brothers, 1921).

p. 47   **No. It isn't** These statements by Trevor Kincaid are from Muriel Guberlet's *The Windows to His World: The Story of Trevor Kincaid* (Palo Alto, CA: Pacific Books, 1975).

p. 47   **Have you noticed** For the ties between the ethos of science and religious asceticism, see David F. Noble's *A World without Women: The Christian Clerical Culture of Western Science* (New York: Knopf, 1992). Darwin's point about a dog speculating on the mind of man is in the letter to Asa Gray in which he mentioned the ichneumons (dated May 22, 1860, and cited above).

p. 48   **Were the taxonomist** Darwin's quip about Nature "telling a direct lie if she can" is related by Raphael Meldola in *Evolution: Darwinian and Spencerian* (Oxford: Clarendon Press, 1910).

p. 48   **Good lord** For Jack London's views of love, see Jack London and Anna Strunsky, *The Kempton-Wace Letters* (New York: MacMillan, 1903).

p. 48   **But Sullivan's** Trevor Kincaid's points regarding the dilemma of the relation between caution (the new brain) and action (the old brain) are based on William Emerson Ritter's *Charles Darwin and the Golden Rule*, edited by Edna Watson Bailey (New York: Storm Publishers, 1954).

Chapter 4

p. 50   **Trevor was no prude** On debates over whether jazz destroyed or improved health, see Russell L. Johnson's "'Disease Is Unrhythmical': Jazz, Health, and Disability in 1920s America," *Health and History* 13 (2011): 13–42. On rules against cheek-to-cheek dancing on campus, see John M. Findlay's "Brides, Brains, and Partisan Politics: Edmond S. Meany, the University of Washington, and State Government, 1889–1939," *Pacific Northwest Quarterly* 99 (2008): 181–93.

p. 52   **I've been offered** On Thomas Hunt Morgan and his fly lab, see Garland Allen's *Thomas Hunt Morgan: The Man and His Science* (Princeton: Princeton University Press, 1979). On the role of women in the early history of genetics, see Marsha L. Richmond's "Women in the Early History of Genetics: William Bateson and the Newnham College Mendelians, 1900–1910," *Isis* 92 (2001): 55–90.

p. 52 **Of a man** On eventual attempts to synthesize lab work and fieldwork, see Robert E. Kohler's *Landscapes and Labscapes: Exploring the Lab-Field Border in Biology* (Chicago: University of Chicago Press, 2002).

p. 53 **Martin hesitated** On Elizabeth Bryant, see Elisabeth Deichmann, "Elizabeth Bangs Bryant," *Psyche* 65 (1958): 3–10. On the status of women in science in the 1920s, in particular their role as illustrators and junior scientific aids rather than curators or professors, see Margaret Rossiter's *Women Scientists in America: Struggles and Strategies to 1940* (Baltimore: Johns Hopkins University Press, 1982).

p. 53 **I suppose** For the application of evolution to the question of women's proper sphere, see Kimberly Hamlin's *From Eve to Evolution: Darwin, Science, and Women's Rights in Gilded Age America* (Chicago: University of Chicago Press, 2014).

p. 56 **Phoebe's interrogation** The quotation is from Patrick Geddes and J. Arthur Thompson's *The Evolution of Sex* (London: Walter Scott, 1899).

p. 56 **Like worshippers at prayer** The description of an evening watching the Nereis worms mating is from Muriel Guberlet's *The Windows to His World: The Story of Trevor Kincaid* (Palo Alto, CA: Pacific Books, 1975).

p. 56 **Too many biologists** Ben Cardiff's comments on the utility of biology to human problems is based on William Emerson Ritter's *The Higher Usefulness of Science* (Boston: RG Badger, 1918).

p. 57 **Oh yes. And that sounds** Ben Cardiff's stance here is based on Albert Wiggam's popular account of the implications of mechanistic biology for human behavior and social control in *The New Decalogue* (Garden City, NY: Garden City, 1925).

p. 58 **Our textbook says** The textbook at issue is Patrick Geddes and J. Arthur Thompson's *The Evolution of Sex* (London: Walter Scott, 1899). For an analysis of Geddes and Thompson's use of biology to justify the status quo, see Cynthia Eagle Russett's *Sexual Science: The Victorian Construction of Womanhood* (Cambridge, MA: Harvard University Press, 2009).

p. 58 **I have no doubt** Ben Cardiff's stance on sex differences is based on biologist David Starr Jordan's *Footnotes to Evolution: A Series of Popular Addresses on the Evolution of Life* (New York: D. Appleton, 1898).

p. 61 **Martin knew better** For the exclusion of women from the Barro Colorado Biological Laboratory, see Pamela Henson's "Invading Arcadia: Women

Scientists in the Field in Latin America, 1900–1950," *The Americas* 58 (2002): 577–600.

p. 61 **She should take** For the roles assigned to women within the eugenics movement, see Amy Sue Bix's "Experiences and Voices of Eugenics Field-Workers: Women's Work in Biology," *Social Studies of Science* 27 (1997): 625–68

p. 62 **Eugene was parroting** On Jacques Loeb, see Heiner Fangerau's "From Mephistopheles to Isaiah: Jacques Loeb, Technical Biology and War," *Social Studies of Science* 39 (2009): 229–56.

p. 63 **Louise smiled too** See University of Washington economist Theresa McMahon's *Women and Economic Evolution; or, The Effects of Industrial Changes upon the Status of Women* (Madison, WI: University of Wisconsin, 1912).

p. 64 **Martin nodded** The views of Sarah's minister-friend are based on Unitarian minister Harold E. B. Speight's sermon *Magic, Science, and Prayer* (Boston: King's Chapel Publications, 1926).

p. 64 **No doubt Freudians** For the contemporary debate over the role of memory in dealing with the human condition and psychotherapeutic treatments, see Robert Armstrong-Jones and Charles Davies-Jones's "Forgetting: Psychological Repression," *British Medical Journal* 1 (1920): 236–37.

p. 66 **He had once heard** For period views of the impact of alcohol on memory, see Sir Victor Horsley and Mary D. Sturge's *Alcohol and the Human Body* (London: MacMillan, 1907) and Richmond Pearson Hobson's *Alcohol and the Human Race* (New York: Fleming H. Revell, 1920)

Chapter 5

p. 69 **The biological station's** A student named Eugene Miller died at the Puget Sound Biological Station when he slipped on the cliff rocks during the summer session of 1924. The account here, however, is imaginary.

p. 69 **Damn rumrunners** On bootlegging in the San Juan Islands, see Rich Mole's *Rum-Runners and Renegades, Whiskey Wars of the Pacific Northwest, 1917–2012* (Victoria, BC: Heritage House, 2013).

p. 73 **Not good enough** The argument that to believe that the world was created by God through natural laws represents a higher view of the Creator may be found in several works, including the concluding chapter of Darwin's *On the Origin of Species* and Reverend Fosdick's "Evolution and Mr. Bryan," *New York Times*, March 12, 1922. For general histories of this view, see Peter Bowler's *Reconciling Science and Religion: The Debate in Early*

Twentieth-Century Britain (Chicago: University of Chicago, 2001) and David N. Livingstone's *Darwin's Forgotten Defenders: The Encounter between Evangelical Theology and Evolutionary Thought* (Grand Rapids, MI: Eerdmans, 1987). Reverend Harrison's diagnosis of the motivation of such claims is based on John W. Porter's *Evolution—a Menace* (Nashville: Southern Baptist Convention, 1922).

p. 73   **I don't blame him** Pete Harrison's confrontation with science as an undergraduate is based on Cyril Harris's *The Religion of Undergraduates* (New York: Charles Scribner's Sons, 1925).

p. 74   **My parents believe** William Jennings Bryan provides this narrative of maternal sacrifice in *In His Image* (New York: Fleming H. Revell, 1922).

p. 74   **Nature, so far as** The Robert Ingersoll quote is from *The Gods: And Other Lectures* (New York: CP Farrell, 1889).

p. 77   **I heard him once** From Billy Sunday's sermon "Spiritual Food for a Hungry World," as excerpted in Edward Larson's *Summer for the Gods: The Scopes Trial and America's Continuing Debate over Science and Religion* (New York: Basic Books, 1997).

p. 77   **He went out** William Jennings Bryan's *In His Image* was based on a series of lectures and published in 1922.

p. 79   **Governor Stevens** Will Sullivan's account of the treaties is based on Cornelius Hanford's *Seattle and Environs* (Seattle: Pioneer Historical Publications, 1924). On the Tulalip Reservation, see Harriette Shelton Dover's *Tulalip, from My Heart: An Autobiographical Account of a Reservation Community* (Seattle: University of Washington Press, 2015). "Scrap of paper" is the phrase the German chancellor used to describe the Treaty of London, in which Britain guaranteed Belgian independence, and which was violated by the German invasion of Belgium in 1914. William Jennings Bryan used the phrase when he insisted that evolution made the Bible a "scrap of paper." See "God and Evolution: Charge That American Teachers of Darwinism 'Make the Bible a Scrap of Paper,'" *New York Times*, February 26, 1922.

Chapter 6

p. 84   **Geneticist William Bateson** The news article is from the *Toronto Globe* for December 29, 1921. For Bateson's speech, see William Bateson's "Evolutionary Faith and Modern Doubts," *Science* 55 (1922): 55–61.

p. 84   **Martin had heard** On William Bateson's 1922 address as the "Sarajevo shot which precipitated war" and scientists' debates regarding what to do in the face of Williams Jennings Bryan's campaign, see Constance Areson Clark's *God—or Gorilla: Images of Evolution in the Jazz Age* (Baltimore: Johns Hopkins University Press, 2008).

p. 84   **For what it's worth** Henry Fairfield Osborn criticized William Bateson's lecture in "William Bateson on Darwinism," *Science* 55 (1922): 194–97. William Jennings Bryan's anti-evolution campaign began two days later.

p. 84   **"Perhaps," he said finally** On criticisms of and alternatives to natural selection, see Peter J. Bowler's *The Eclipse of Darwinism: Anti-Darwinian Evolution Theories in the Decades around 1900* (Baltimore: Johns Hopkins University Press, 1992).

p. 85   **Some believe** Reverend Harrison's stance (including his definition of science) is based on William Jennings Bryan's "The Fundamentals," *Forum* 70 (1923): 1675–1680; "Bryan Raps Colleges for Lack of Religion: Names Columbia in Group He Says Is Turning Out Infidels and Skeptics," *New York Times*, November 26, 1921; and the secretary of the Anti-Evolution League T. T. Martin's *Hell and the High Schools: Christ or Evolution, Which?* (Kansas City: Western Baptist, 1923) and his "Three Fatal Teachings of President Poteat of Wake Forest College," *Western Recorder* (February 5, 1920): 4–5. Other examples of influential anti-evolutionist literature from the period include John W. Porter's *Evolution—a Menace* (Nashville: Southern Baptist Convention, 1922) and Alfred Watterson McCann's *God—or Gorilla: How the Monkey Theory of Evolution Exposes Its Own Methods, Refutes Its Own Principles, Denies Its Own Inferences, Disproves Its Own Case* (New York: Devin-Adair, 1922).

p. 86   **But there is a tenet of faith** For natural law as a tenet of faith in science, see Edward E. Slosson's "The Faith of the Scientist," *Scientific Monthly* 17 (1923): 510–12.

p. 86   **And if science is knowledge** On fundamentalists' objection to particular types of science (especially evolution) rather than science as a whole (when defined a certain way), see George Marsden's "Fundamentalism as an American Phenomenon, a Comparison with English Evangelicalism," *Church History* 46 (1977): 215–32.

p. 86   **Martin said nothing** The exchange regarding the heron and the fish is based on Robert Ingersoll's *The Gods: And Other Lectures* (New York: CP Farrell, 1889).

p. 88   **But that's just it** The list of fundamentals is based on that given by William Bell Riley in "The Faith of the Fundamentalists," *Current History* 26 (1927): 434–36.

p. 89   **The individual agnostic** On the use of Darwinian theory by German militarists to justify the invasion of Belgium, see entomologist Vernon Kellogg's *Headquarters Nights: A Record of Conversations and Experiences at the Headquarters of the German Army in France and Belgium* (Boston: Atlantic Monthly Press, 1917). This book had an important impact on William Jennings Bryan's campaign.

p. 89   **Martin cringed** Clarence Darrow's closing arguments in the Leopold and Loeb trial may be found in Simone Payment's *The Trial of Leopold and Loeb: A Primary Source Account* (New York: Rosen, 2003).

p. 90   **I doubt that** Scientists countered the claim that their science courses inspired Leopold and Loeb in "Science Did Not Inspire Chicago Murder, University Records Show," *Science News-Letter* 4 (1924): 1–2.

p. 95   **Martin had smiled** See Albert Wiggam's "The New Decalogue of Science: An Open Letter from the Biologist to the Statesman," *Century Magazine* 103 (1922): 643–50. For the history of eugenics in the United States, see Diane Paul's *Controlling Heredity: 1865 to the Present* (Amherst, NY: Prometheus Press, 1995), Daniel Kevles's *In the Name of Eugenics: Genetics and the Uses of Human Heredity* (Cambridge, MA: Harvard University Press, 1995), Mark Largent's *Breeding Contempt: The History of Coerced Sterilization in the United States* (New Brunswick, NJ: Rutgers University Press, 2011), and Alexandra Minna Stern's *Eugenic Nation: Faults and Frontiers of Better Breeding in Modern America* (Berkeley: University of California Press, 2016).

p. 95   **Martin took the magazine** Support for Albert Wiggam's campaign among biologists can be found at "Letters to the Editor," *Century Magazine* 104 (1922): 973.

p. 96   **Martin had never paid much mind** Books on eugenics by prominent biologists included, from Stanford University, David Starr Jordan's *The Human Harvest: A Study of the Decay of the Races through the Survival of the Unfit* (Boston: American Unitarian Association, 1907); from the Eugenics Record Office, Charles Davenport's *Eugenics: The Science of Human Improvement by Better Breeding* (New York: Henry Holt, 1910) and *Heredity in Relation to Eugenics* (New York: Henry Holt, 1911); from Goucher College, William Kellicott's *The Social Direction of Human Evolution: An Outline of the Science of Eugenics* (New York: D. Appleton, 1911); and from Harvard, Edward East's *Mankind at the Crossroads* (New York: Charles Scribner's Sons, 1923) and William Castle's *Genetics and Eugenics: A Text-book for Students of Biology and a Reference Book for Animal and Plant Breeder* (Cambridge, MA: Harvard University Press, 1924). For Henry Fairfield Osborn's call for eugenics, see his "The Second International Congress of Eugenics Address of Welcome," *Science* 54 (1921): 311–13.

p. 97   **Martin hesitated** By 1928 eugenics was offered in 376 college courses across the country. On the popularity of eugenics at universities, see "Eugenics in the Colleges," *Journal of Heredity* 5 (1914): 186. Biology professor James R. Slater taught the College of Puget Sound's course from 1920 to 1951. Trevor Kincaid taught the University of Washington's course from 1914 to 1949.

Chapter 7

p. 100   **Mental Hygiene and Eugenics** The course description is from the *College of Puget Sound Bulletin, Catalog Edition*, vol. 16, no. 2 (Tacoma, WA: College of Puget Sound, April 1924).

p. 101   **Well, it's the kind** On the essay contest sponsored by the College of Puget Sound and Pierce County Social Hygiene Society, see the *Social Hygiene Bulletin* 8 (1921): 8.

p. 101   **What are some of the characteristics** The list comes from a record of Professor James R. Slater's exam questions found in the University of Puget Sound Archives and Special Collections, RG 10.05 Biology Department, box 2.

p. 102   **Martin might have thought** For Havelock Ellis on clothing and desire, see his "Sexual Education and Nakedness," *American Journal of Psychology* 20 (1909): 297–317.

p. 104   **It is interesting** The quotation is from Charles Darwin's *On the Origin of Species by Means of Natural Selection, or, the Preservation of the Favoured Races in the Struggle for Life*, 6th ed. (London: J. Murray, 1876).

p. 105   **That's strange** One can trace Darwin's changes in each edition by browsing *The Preservation of Favoured Traces* at www.benfry.com/traces. On Darwin's "God Talk," see David Kohn's "Darwin's Ambiguity: The Secularization of Biological Meaning," *British Journal for the History of Science* 22 (1989): 215–39.

p. 105   **That's a minor tragedy** On John Gulick, see Addison Gulick's *Evolutionist and Missionary, John Thomas Gulick: Portrayed through Documents and Discussion* (Chicago: University of Chicago Press, 1932). On the reconciliations developed by American theologians, see Cynthia Russett's *Darwin in America: The Intellectual Response, 1865–1912* (San Francisco: Freeman, 1976), David N. Livingstone's *Darwin's Forgotten Defenders: The Encounter between*

Evangelical Theology and Evolutionary Thought (Grand Rapids, MI: Eerdmans, 1987), and James Moore's *The Post-Darwinian Controversies: A Study of the Protestant Struggle to Come to Terms with Darwin in Great Britain and America, 1870–1900* (Cambridge: Cambridge University Press, 1979).

p. 107 **No, I don't** For criticisms of those who opposed evolution and Christianity on these grounds, see Leighton Parks's *What Is Modernism?* (New York: Scribner's, 1924), Edward Mortimer Chapman's *A Modernist and His Creed* (Boston: Houghton Mifflin, 1926), or any of the writings of Harry Emerson Fosdick and Shailer Mathews, including Fosdick's "Shall the Fundamentalists Win?," *Christian Work* 102 (1922): 716–72, and "Evolution and Mr. Bryan," *New York Times*, March 12, 1922.

p. 113 **Who was that anthropologist** The discussion of European versus indigenous practices in naming the landscape is based on anthropologist T. T. Waterman's "The Geographical Names Used by the Indians of the Pacific Coast," *Geographical Review* 12 (1922): 174–94.

p. 113 **Some rich city ladies** Henry Sicade spoke to the North West Real Estate Association Convention in Rainier National Park in 1921. He saw the dispute as a distraction from the more important issue of the dispossession of reservation lands to settlers. Sicade's view, and his reply to Mrs. M. G. Mitchell when she asked him to organize an appeal from the Puyallup, can be found in Lisa Blee's "Mount Rainier and Indian Economies of Place, 1850–1925," *Western Historical Quarterly* 40 (2009): 419–43. Sicade's own account of the origin of the mountain's name can be found in "Aboriginal Nomenclature," *Mazama* 5 (1918): 251–54.

p. 114 **Rebecca looked at him** Rebecca's comments on the "great Spirit, Creator of all things," are based on Henry Sicade's "The Indian's Side of the Story," address to the Research Club of Tacoma, April 10, 1917, reprinted in *Building a State, Washington: 1889–1939*, edited by Charles Miles and O. B. Sperlin (Olympia: Washington State Historical Society, 1940), 490–503.

p. 115 **Martin suspected** On marriages between settlers and Native Americans, see Katrina Jagodinsky's *Legal Codes and Talking Trees: Indigenous Women's Sovereignty in the Sonoran and Puget Sound Borderlands 1854–1946* (New Haven, CT: Yale University Press, 2016).

p. 115 **My grandparents married** On the fact traditional incentives to marry outside one's group became a liability in the context of European racial thinking, see Alexandra Harmon's "Lines in Sand: Shifting Boundaries between Indians and Non-Indians in the Puget Sound Region," *Western Historical Quarterly* 26 (1995): 429–53.

p. 116 **Those still aren't on the books** On the history of antimiscegenation laws, see Paul Lawrence Farber's *Mixing Races: From Scientific Racism to Modern Evolutionary Ideas* (Baltimore: Johns Hopkins University Press, 2011). The founder of the Tacoma NAACP, Dr. Nettie Asberry, successfully fought against such laws in Washington State. See Antoinette Broussard's "Nettie Craig Asberry: A Pillar of Tacoma's African American Community," *Columbia* 19 (2005): 3–6.

p. 116 **Her eldest boy** On the Cushman Indian School, see "Hard Lessons in America: Henry Sicade's History of Puyallup Indian School, 1860–1920," edited by Cary C. Collins, *Columbia* 14 (2000–2001), and Charles Roberts's "The Cushman Indian Trades School and World War I," *American Indian Quarterly* 11 (1987): 221–39. Parts of Rebecca's story were inspired by the life of Rebecca Lena Graham, as told in Katrina Jagodinsky's *Legal Codes and Talking Trees: Indigenous Women's Sovereignty in the Sonoran and Puget Sound Borderlands, 1854–1946* (New Haven, CT: Yale University Press, 2016).

Chapter 8

p. 119 **These monkeys** Quotations attributed to Charles Darwin's *The Descent of Man* are from the second edition published by John Murray in London in 1874.

p. 120 **Bryan argues** The comparison between the use of natural law to explain individual versus species development was often made by those defending evolution against charges of impiety. See, for example, Robert Chambers's *Vestiges of the Natural History of Creation* (London: Churchill, 1844).

p. 124 **Then he didn't think** For an analysis of Darwin's explanation of the evolution of the moral sense, see Robert Richards's "Darwin on Mind, Morals, and Emotions," in *The Cambridge Companion to Darwin*, edited by J. Hodge and G. Radick (Cambridge: Cambridge University Press, 2003). The use of Darwin's theory to explain the human mind formed a stumbling block for many of his allies, including Charles Lyell, Alfred Russel Wallace, and Frances Power Cobbe. See *Sir Charles Lyell's Scientific Journals on the Species Question*, edited by Leonard G. Wilson (New Haven, CT: Yale University Press, 1970), the final chapter of Wallace's *Darwinism* (London: Macmillan, 1890), and Cobbe's *Darwinism in Morals and Other Essays* (London: Williams and Norgate, 1872).

p. 127 **He was a naturalist** For an analysis of the racial language in *The Descent* *of Man*, its origins and influence, see John Greene's "Darwin as a Social Evolutionist," *Journal of the History of Biology* 10 (1977): 1–27, and Matthew Day's "Godless Savages and Superstitious Dogs: Charles Darwin, Imperial Ethnography, and the Problem of Human Uniqueness," *Journal of the History of Ideas* 6 (2008): 49–70.

p. 129 **But isn't it the missionaries** On the debate over the policies of the Bureau of Indian Affairs, and the attitudes of both missionaries and reformers like John Collier, see David W. Daily's *Battle for the BIA: G. E. E. Lindquist and the Missionary Crusade against John Collier* (Tucson: University of Arizona Press, 2004)

p. 129 **You think that makes up** Adrian Desmond and James Moore, in *Darwin's Sacred Cause: How a Hatred of Slavery Shaped Darwin's Views on Human Evolution* (Boston: Houghton Mifflin Harcourt, 2009), note how the Anglican reverend and Darwinian Charles Kingsley used natural selection to argue that "the 'lowly' races were Providentially doomed and that the whites would sweep out all before them to usher in God's Kingdom" (318). For the claim that diseases like smallpox prepared the Pacific Northwest for white settlers, see Leslie M. Scott's "Indian Diseases as Aids to Pacific Northwest Settlement," *Oregon Historical Quarterly* 29 (1928): 144–61.

p. 129 **Two of Rebecca's cousins** For Commissioner Darwin and the use of "natural law" to justify policies toward the indigenous peoples of the Pacific Northwest, see Charles M. Buchanan, "Rights of the Puget Sound Indians to Game and Fish," *Washington Historical Quarterly* 6 (1915): 109–18. The case in which Commissioner Darwin arrested two Lummi tribe members (Dan Ross and Patrick George) actually took place in 1913.

p. 130 **People complain** On the policies of the Bureau of Indian Affairs as the outcome of the belief in the "survival of the fittest," see David Wallace Adams, *Education for Extinction: American Indians and the Boarding School Experience, 1875–1928* (Lawrence: University Press of Kansas, 1995).

Chapter 9

p. 135 **Martin had at first** For an example of Havelock Ellis's defense of eugenics on the grounds it would meliorate suffering, see his *Little Essays of Love and Virtue* (London: A.&C. Black, 1922).

p. 135 **Now, a stack of books** See G. H. Parker, "The Eugenics Movement as a Public Service," *Science* 41 (1915): 342–47.

p. 136 **This was applied biology** Based on the book form of Albert Wiggam's essay, *The New Decalogue* (Garden City, NY: Garden City, 1925).

p. 138 **Perhaps the most valuable** See Anna Botsford Comstock's *Handbook of Nature Study* (Ithaca, NY: Comstock, 1911).

p. 138 **To his very core** C. C. Little's stance is outlined in his "The Relation between Research in Human Heredity and Experimental Genetics," *Scientific Monthly* 14 (1922): 401–14. On Edwin Conklin, see Kathy J. Cooke's "Duty or Dream? Edwin G. Conklin's Critique of Eugenics and Support for American Individualism," *Journal of the History of Biology* 35 (2002): 365–84. The point about dice was made by Dr. C. A. Mercier during a discussion following Francis Galton's paper "Eugenics: Its Definition, Scope, and Aims" in the *American Journal of Sociology* 10 (1905): 1–25.

p. 142 **Cushman Indian School** See "Hard Lessons in America: Henry Sicade's History of Puyallup Indian School, 1860–1920," edited by Cary C. Collins, *Columbia* 14 (2000–2001).

p. 143 **Alien Land Act** On the 1921 Alien Land Act and Japanese farmers in Washington, see S. Frank Miyamoto's "The Japanese Minority in the Pacific Northwest," *Pacific Northwest Quarterly* 54 (1963): 143–49.

p. 144 **They were going** On the role of Trevor Kincaid, Emy Tsukimoto, and Joe Miyagi in the history of the oyster industry, see Kathleen Whalen Fry, "Transforming the Tidelands: Japanese Labor in Washington's Oystering Communities before 1942," *Pacific Northwest Quarterly* 102 (2001): 132–43, and E. N. Steele's *The Immigrant Oyster (*Ostrea gigas*) Now Known as the Pacific Oyster* (Seattle: Pacific Coast Oyster Growers Association, 1964). Kincaid's actual stance on these moves is unknown.

p. 147 **Reade asks** See Winwood Reade's *The Martyrdom of Man* (New York: Asa K. Butts, 1874). For an example of the claim that Darwin triumphed because Victorians were tired of orthodox explanations of suffering, see Bernard Shaw's *Back to Methuselah: A Metabiological Pentateuch* (New York: Brentano's, 1921). For an example of the appeal of evolution given the problem of suffering, see the private notebooks of Charles Lyell in *Sir Charles Lyell's Scientific Journals on the Species Question*, edited by Leonard G. Wilson (New Haven, CT: Yale University Press, 1970).

p. 149 **Martin shook his head** For Sam Henshaw's attitude toward applied work and chemicals, see Robert J. Spear's *The Great Gypsy Moth War: The History of the First Campaign in Massachusetts to Eradicate the Gypsy Moth, 1890–1901* (Amherst: University of Massachusetts Press, 2005).

p. 151 **He and Sarah had had an argument** The stance that new ideas of prayer and God's goodness are necessary is based on Rev. Harold E. B. Speight's sermon *Magic, Science, and Prayer* (Boston: King's Chapel Publications, 1926). For other examples of the claim science is provided by God to alleviate suffering, see Kristin Johnson's "Furnishing the Skill Which Can Save the Child: Diphtheria, Germ Theory, and Theodicy," *Zygon* 52 (2017): 296–322.

Chapter 10

p. 153 James Slater had lent him See Paul Popenoe's *Applied Eugenics* (New York: Macmillan, 1918) and Herbert Eugene Walter's *Genetics: An Introduction to the Study of Heredity* (New York: Macmillan, 1913). Both these books are cited in the thesis of James Legg (Slater's student).

p. 154 Then Martin thought On the role of Edward Murray East and Edwin G. Conklin in the debate over miscegenation, see Bentley Glass's "Geneticists Embattled: Their Stand against Rampant Eugenics and Racism in America during the 1920s and 1930s," *Proceedings of the American Philosophical Society* 180 (1986): 130–54.

p. 157 I know some of your colleagues Frank Gray's point about the utility of an occasional reminder that men were warriors once is from Vernon Kellogg's *Beyond War: A Chapter in the Natural History of Man* (New York: Henry Holt, 1912).

p. 158 Martin panicked For H. L. Mencken's use of evolution to oppose socialism, see Mencken and Robert Rives LeMonte's *Men versus the Man: A Correspondence between Robert Rives La Monte, a Socialist, and H. L. Menken, an Individualist* (New York: Henry Holt, 1910).

p. 159 That doesn't matter On the importance of Darwin to socialists, see Mark A. Pittenger's *American Socialists and Evolutionary Thought, 1870–1920* (Madison: University of Wisconsin Press, 1993).

p. 160 Everyone picks and chooses On the rise of biological explanations of poverty and crime in the nineteenth century, see Daniel Pick's *Faces of Degeneration: A European Disorder c. 1848–1918* (Cambridge: Cambridge University Press, 1989) and Robert A. Nye's "The Rise and Fall of the Eugenics Empire: Recent Perspectives on the Impact of Biomedical Thought in Modern Society," *Historical Journal* 36 (1993): 687–700. For Kropotkin's views, see his *Mutual Aid: A Factor in Evolution* (New York: McClure Phillips, 1902) and Stephen Jay Gould's "Kropotkin Was No Crackpot," *Natural History* 97 (1988): 12–21.

p. 166 **Frank guided him** On the difficulty of maintaining teachers at Cushman Indian School because they were easily tempted by the "Bolsheviki spirit" in Tacoma, see page 450 of *Indian Appropriation Bill: Hearings before a Subcommittee of the Committee on Indian Affairs of the House of Representatives* (Washington, DC: Washington Government Printing Office, 1920).

p. 166 **Martin was soon forgotten** For a description of offices of the Industrial Workers of the World and a reading list of Seattle's socialists, see Robert L. Tyler's *Rebel in the Woods: The I.W.W. in the Pacific Northwest* (Eugene, OR: University of Oregon Press, 1967).

p. 167 **Frank shook his head** On the hopes that Seattle would be the center of the revolution, see Kristofer Allerfeldt's *Race, Radicalism, Religion, and Restriction* (New York: Praeger, 2003). On Harry Ault and criticisms of his watered-down socialism, see John C. Putman's *Class and Gender Politics in Progressive-Era Seattle* (Reno: University of Nevada Press, 2008).

p. 167 **One wouldn't know that** Washington State representative Albert Johnson campaigned successfully, with the help of the Eugenics Record Office's Harry Laughlin (the "Expert Eugenical Agent" of the House Committee on Immigration and Naturalization), for an immigration quota of 2 percent of each group's population (based on the 1890 census). Those from the "Asiatic Barred Zone" were barred entirely. The Immigration Act was passed in April 1924. See Kristofer Allerfeldt's "'And We Got Here First': Albert Johnson, National Origins and Self-Interest in the Immigration Debate of the 1920s," *Journal of Contemporary History* 45 (2010): 7–26. For David Starr Jordan's stance on immigration laws, see his "Should Present Restriction Laws be Modified?" *Congressional Digest* 2 (1923): 304–5. The *New York Times* piece was titled "Biologist Supports Curb on Immigrants," *New York Times*, April 6, 1924. On its author, Ivey F. Lewis, see Gregory Michael Door's "Assuring America's Place in the Sun: Ivey Foreman Lewis and the Teaching of Eugenics at the University of Virginia, 1915–1953," *Journal of Southern History* 66 (2000): 257–96.

p. 168 **He had had to follow** For Herbert Spencer Jennings's criticisms of eugenics and his testimony at the immigration hearings, see Kenneth Ludmerer's "Genetics, Eugenics, and the Immigration Restriction Act of 1924," *Bulletin of the History of Medicine* 46 (1972): 59–81, and Elazar Barkan's "Reevaluating Progressive Eugenics: Herbert Spencer Jennings and the 1924 Immigration Legislation," *Journal of the History of Biology* 24 (1991): 91–112.

p. 168 Didn't do any good Harry Ault's stance that men were cloaking race prejudice with biology is based on John Langdon-Davies's critique of "race fiends" in *The New Age of Faith* (New York: Viking Press, 1925).

p. 168 I said my stance Harry Ault's testimony is given in *Japanese Immigration: Hearings before the Committee on Immigration and Naturalization, House of Representatives, Sixty-Sixth Congress, Second Session, July 12, 13, and 14, 1920* (Washington, DC: Government Printing Office, 1921). The "laws of nature" were cited by Senator Ben D. Phelan in his remarks against assimilation.

p. 169 Surely men denounce For the role of biblical defenses of inequality and segregation on the grounds that racial barriers were created by God (in the context of African American elites' support for evolutionary visions of the unity of man), see Jeffrey P. Moran's "Reading Race into the Scopes Trial: African American Elites, Science, and Fundamentalism," *Journal of American History* 90 (2003): 891–911. Also see Fay Botham's *Almighty God Created the Races: Christianity, Interracial Marriage, and American Law* (Chapel Hill: University of North Carolina Press, 2009).

p. 169 Harry nodded For contemporary criticisms of white supremacists' claims that science was on their side, see Hendrik Willem Van Loon's "Our Nordic Myth-Makers," *The Nation* 120 (1925): 349–50

p. 169 Frank stood For the IWW's stance regarding the relationship between World War I and capitalism, see Kristofer Allerfeldt's *Race, Radicalism, Religion, and Restriction* (New York: Praeger, 2003) and Jeffrey A. Johnson's *They're All Red Out Here: Seattle Politics in the Pacific Northwest, 1895–1925* (Norman: University of Oklahoma, 2008).

p. 171 Comrades, to accept Frank Gray's critique of American socialists who viewed progress as inevitable is based on Mark A. Pittenger's *American Socialists and Evolutionary Thought, 1870–1920* (Madison: University of Wisconsin Press, 1993). Frank Gray's stance on science and antiracism is based on NAACP cofounder William English Walling and the thought of a group Pittenger refers to as the "New Intellectual" socialists. See Walling's "Science and Human Brotherhood," *The Independent* 66 (1909): 1318–27. On the antiracism stance of the IWW, see Kristofer Allerfeldt's *Race, Radicalism, Religion, and Restriction* (New York: Praeger, 2003).

Chapter 12

p. 177 All against the rules On the response of Bureau of Indian Affairs agents to the Shaker religion, see Susan Neylan's "Shaking Up Christianity: The

Indian Shaker Church in the Canada-US Pacific Northwest," *Journal of Religion* 91 (2011): 188–222.

p. 179 **He read several pages** On Lucy Maynard Salmon, see Chara Haeussler Bohan's *Go to the Sources: Lucy Maynard Salmon and the Teaching of History* (New York: P. Lang, 2004). Salmon's books include *The Newspaper and the Historian* (Oxford: Oxford University Press, 1923), *Domestic Service* (New York: Macmillan, 1901), and *Progress in the Household* (Boston: Houghton Mifflin, 1906).

p. 183 **She glanced up at him** Charlotte Perkins Gilman's essay "Do Women Dress to Please Men" appeared in *Century Magazine* 103 (1922): 651–59.

p. 186 **He pulled a paper** Trevor Kincaid's lecture was advertised in *Seattle Labor College: Program of Classes and Lectures, Third Term—Year 1922–23* (Seattle: Seattle Labor College, 1923).

p. 187 **No wonder you** On President Suzzallo's trouble in the face of his support for the moderate labor movement, and his use of a "secret service" to keep tabs on faculty radicalism, see Patrick Farrell's "The Campus Kaiser: Henry Suzzallo, Militarism, the University of Washington and Labor Politics from 1915–1920" (1999), available online through the Seattle General Strike Project, part of the Great Depression in Washington State Project, hosted at https://depts.washington.edu/labhist/.

p. 188 **Martin sat at his office desk** This "loss of the collection" story is fiction, but fraternizing with the IWW could indeed get faculty in trouble. President Suzzallo terminated the contract of English teacher Miss Gregg after she pretended to sympathize with two undercover detectives posing as Wobblies at a Seattle Labor Council meeting. See *Lone Voyagers: Academic Women in Coeducational Institutions, 1870–1937*, edited by Geraldine Jonçich Clifford (New York: Feminist Press, 1989).

Chapter 13

p. 193 **Martin hated** On W. B. Riley, see William Vance Trollinger Jr.'s *God's Empire: William Bell Riley and Midwestern Fundamentalism* (Madison: University of Wisconsin Press, 1991). For a sample of John Roach Straton's speeches, see "John Roach Straton Condemns Evolution as Degenerate Cult," *Harvard Crimson* (October 16, 1925).

p. 193 **Then Martin had to listen** For the stances of Alonzo Baker and Maynard Shipley in the radio debates, see Edward Larson's *Summer for the Gods: The Scopes Trial and America's Continuing Debate over Science and Religion* (New

York: Basic Books, 1997) and Alonzo L. Baker's "The San Francisco Evolution Debates, June 13–14, 1925," *Adventist Heritage* 2 (1975): 23–31.

p. 194 **Reports on developments** For Henry Fairfield Osborn's reply to John Roach Straton, see Osborn's 1924 address to Columbia University Assembly, printed in his book *Evolution and Religion in Education* (New York: Charles Scribner's Sons, 1926).

p. 196 **So she sat down again** The medieval bestiary tales are from Thomas H. White's *The Book of Beasts* (New York: G. P. Putnam's Sons, 1954) and volume 1 of A. C. Crombie's *Medieval and Early Modern Science* (Garden City, NY: Doubleday, 1959).

p. 197 **But I know men** The contemporary naturalist William Beebe would leave the room when colleagues used biology to justify political and social ideologies. See Carol Grant Gould's *The Remarkable Life of William Beebe: Explorer and Naturalist* (Washington, DC: Island Press, 2004).

p. 201 **He's assuming** On the influence of Darwin's motivations and Victorian "separate of spheres" ideology on his work, see Evelleen Richards's "Darwin and the Descent of Woman," in *The Wider Domain of Evolutionary Thought* (Dordrecht: Springer, 1983).

p. 202 **Some thought** For Elizabeth Cady Stanton's argument that Darwin's theory would emancipate women, see Stanton et. al.'s *The Woman's Bible* (New York: European, 1895).

p. 203 **She smiled again** On the higher critics and their response to Darwin's work, see John Hedley Brooke's *Science and Religion: Historical Perspectives* (Cambridge: Cambridge University Press, 1991).

p. 203 **I did** The claim about medieval theologians' understanding of insects is in Karl Jordan's "The President's Address," *Proceedings of the Royal Entomological Society of London* 5 (1931): 128–42. For an overview of changing concepts of God associated with the rise of rationalism and science, see Frank Turner's *Without God, Without Creed: The Origins of Unbelief in America* (Baltimore: Johns Hopkins University Press, 1985). For orthodox Christians who reconciled Darwin's theory with belief in a fallen creation, see James R. Moore's *The Post-Darwinian Controversies: A Study of the Protestant Struggle to Come to Terms with Darwin in Great Britain and America, 1870–1900* (Cambridge: Cambridge University Press, 1981).

p. 206 **You remind me** The rule that man cannot know the mind of God, sometimes called the divine omnipotence argument, played an important role in the Galileo trial. See Maurice Finocchiaro's *The Trial of Galileo: Essential Documents* (Indianapolis, IN: Hackett, 2014).

p. 206 **You make it sound** On the decreasing status of taxonomists, see Kristin Johnson's "Natural History as Stamp Collecting: A Brief History," *Archives of Natural History* 34 (2007): 244–58.

p. 207 **No. I don't** For a contemporary defense of natural history on the grounds it cultivates a habit of mind, see the works of California biologist William Emerson Ritter.

p. 209 **People have a hard time** Helen Gray's point about the impression given by the available records is based on the review of Eileen Power's *Medieval English Nunneries* by Bertha Haven Putnam in *American Historical Review* 29 (1924): 538–39.

p. 209 **Martin was sitting** The first example is from Herbert Eugene Walter's *Genetics: An Introduction to the Study of Heredity* (New York: Macmillan, 1913). The second is from Edwin Conklin's *Heredity and Environment in the Development of Men*, 2nd ed. (Princeton: Princeton University Press, 1920).

p. 210 **It's often told** Zoologist Edwin G. Conklin told as truth the story of the medieval belief in a flat earth in "Bryan and Evolution Why His Statements Are Erroneous and Misleading—Theology Amusing If Not Pathetic," *New York Times* March 5, 1922. On the use of the "Flat Earth Error" to bolster nineteenth-century assumptions regarding progress, see Jeffrey Burton Russell's *Inventing the Flat Earth: Columbus and Modern Historians* (New York: Praeger, 1991).

p. 211 **You're both in good company** On Andrew Dickson White's version of history, see Glenn C. Altschuler's "From Religion to Ethics: Andrew D. White and the Dilemma of a Christian Rationalist," *Church History* 47 (1978): 308–24, and Richard Schaefer's "Andrew Dickson White and the History of a Religious Future," *Zygon* 50 (2015): 7–27.

p. 212 **She nodded** On George Lincoln Burr and Andrew Dickson White's partnership and their differences, see Henry Guerlac's "George Lincoln Burr," *Isis* 35 (1944): 147–52.

Chapter 14

p. 214 **Martin had never been so aware** On the president of Harvard's opposition to hiring women, see Jenna Tonn's "Extralaboratory Life: Gender Politics and Experimental Biology at Radcliffe College, 1894–1910," *Gender & History* 29 (2017): 329–58. For biologists' discussion of women's roles, see Susan Sleeth Mosedale's "Science Corrupted: Victorian Biologists Consider 'The Woman Question,'" *Journal of the History of Biology* 11 (1978): 1–55.

p. 214 **He knew all this because** For examples of the eugenic family studies, see *White Trash: The Eugenic Family Studies, 1877–1919*, edited by Nicole Han Rafter (Boston: Northeastern University Press, 1988). The study described here is Wilhelmine Key's *Heredity and Social Fitness: A Study of Differential Mating in a Pennsylvania Family* (Washington, DC: Carnegie Institution, 1920).

p. 216 **We know enough** For the views described here on eugenics, female education, and female psychology, see G. Stanley Hall's "Eugenics: Its Ideals and What It Is Going to Do," *Religious Education* 6 (1911): 152–59, and *Adolescence: Its Psychology and Its Relations to Physiology, Anthropology, Sociology, Sex, Crime, Religion and Education* (New York: D. Appleton, 1915).

p. 218 **Plenty of upstanding** James Slater's point that there were orthodox precedents for eugenics is based on G. Stanley Hall's "Eugenics: Its Ideals and What It Is Going to Do," *Religious Education* 6 (1911): 152–59, and Christine Rosen's *Preaching Eugenics: Religious Leaders and the American Eugenics Movement* (Oxford: Oxford University Press, 2004). His comments on eugenics and religion are based on the sermons and writings of Rev. Kenneth C. MacArthur and Dr. Harry F. Ward, as quoted in Rosen's *Preaching Eugenics* and contained in the first volume of the journal *Eugenics* (1928), an issue devoted to cooperation between eugenicists (the term *eugenists* was used in the 1920s) and ministers. The statement on what God desires is from Slater's student James S. Legg's thesis "Eugenic Sterilization" (College of Puget Sound, 1947).

p. 219 **Three hundred and fifty** For the number of eugenics courses in 1928, see Hamilton Cravens's *The Triumph of Evolution: American Scientists and the Heredity-Environment Controversy, 1900–1941* (Philadelphia: University of Pennsylvania Press, 1978).

p. 219 **I understand, Sullivan** The point about doing good (by applying eugenics) even when our knowledge is limited was made by Darwin's son, Leonard Darwin, and quoted in E. S. Gosney and Paul Popenoe's *Sterilization for Human Betterment: A Summary of Results of 6,000 Operations in California, 1909–1929* (New York: MacMillan, 1931), later cited by James Slater's student James Legg.

p. 220 **This nation lost** James Slater's concerns regarding the "welfare of the race" in the wake of World War I are based on those of David Starr Jordan, whose *War and the Breed: The Relation of War to the Downfall of Nations* (Boston: Beacon Press, 1915) was in turn used by pacifists like Will Irwin in *The Next War: An Appeal to Common Sense* (New York: E. P. Dutton, 1921).

p. 220 But Slater, it turned out On the tensions in *The Descent of Man* concerning eugenic thinking, see Diane Paul's "Darwin, Social Darwinism and Eugenics," in *The Cambridge Companion to Darwin*, edited by Jonathan Hodge and Gregory Radick (Cambridge: Cambridge University Press, 2009), and John Greene's "Darwin as a Social Evolutionist," *Journal of the History of Biology* 10 (1977): 1–27.

p. 221 He sat back The point about sacrifice was made by James Slater's student James Legg in his thesis "Eugenic Sterilization" (College of Puget Sound, 1947).

p. 223 The war. Yes On the impact of World War I on taxonomy and the colonial secretary of Jamaica's opposition to collecting, see Kristin Johnson's *Ordering Life: Karl Jordan and the Naturalist Tradition* (Baltimore: Johns Hopkins University Press, 2012).

p. 225 She shook her head For an emphasis on national interests as the cause of World War I, see Edward Raymond Turner's "The Causes of the Great War," *American Political Science Review* 9 (1915): 16–35. For a defense of Nietzsche in the face of accusations he was to blame for World War I, see William Mackintire Salter's "Nietzsche and the War," *International Journal of Ethics* 27 (1917): 357–79. On the impact of World War I on some historians' commitment to their discipline, see Clarence Walworth Alvord's "Musings of an Inebriated Historian," *American Mercury* 5 (1925): 434–41.

p. 225 We can go you one better For the debates over whether war or peace was natural and over biology as a factor in causing World War I, see Paul Crook's *Darwinism, War and History: The Debate over the Biology of War from the 'Origin of Species' to the First World War* (Cambridge: Cambridge University Press, 1994).

Chapter 15

p. 230 He'd read plenty of explanations On love as a mix of chemical reactions, see Jack London and Anna Strunsky's *The Kempton-Wace Letters* (New York: MacMillan, 1903).

p. 231 Have you ever The quotation is from William Jennings Bryan's *In His Image* (New York: Fleming H. Revell, 1922).

p. 231 J. T. Scopes This is from the article "Arrest under Evolution Law," *Nashville Banner*, May 6, 1925, as quoted in Edward Larson's *Summer for the Gods: The Scopes Trial and America's Continuing Debate over Science and Religion* (New York: Basic Books, 1997).

p. 232 SCOPES WILL FIGHT The two headlines are from the *New York Times*, May 17 and 26, 1925.

p. 233 I don't know On socialist versions of eugenics, see Diane Paul's "Eugenics and the Left," *Journal of the History of Ideas* 45 (1984): 567–90.

p. 233 Martin couldn't ignore anything The statement about the Nineteenth Amendment is quoted in Jeffrey P. Moran's *American Genesis: The Evolution Controversies from Scopes to Creation Science* (Oxford: Oxford University Press, 2012).

p. 235 It's not your fault This summary of the history of the Duwamish people is based on Lea Stickley Dicker's "The Tribe That Wouldn't Die," *Seattle Met*, March 2009, and Katrina Jagodinsky's *Legal Codes and Talking Trees: Indigenous Women's Sovereignty in the Sonoran and Puget Sound Borderlands, 1854–1946* (New Haven, CT: Yale University Press, 2016). On the continuing struggle of the Duwamish for federal recognition visit www.duwamishtribe.org.

p. 238 Dad's been reading See Scudder Klyce's *The Sins of Science* (Boston: Marshall Jones, 1925).

p. 242 There are fire-stones These passages are from Thomas H. White's *The Book of Beasts* (New York: G. P. Putnam's Sons, 1954).

p. 243 Had I lived a couple Thomas Henry Huxley's letter to Reverend Kingsley may be found in volume 1 of *Life and Letters of Thomas Henry Huxley*, edited by Leonard Huxley (New York: D. Appleton, 1901).

p. 245 The memory of The quotation is from William Jennings Bryan's *In His Image* (New York: Fleming H. Revell, 1922).

Chapter 16

p. 246 Trevor groaned The AAAS Council's resolution defending evolution as both true and a great good can be found in the *Bulletin of the Association of University Professors* 9, no. 3 (1923): 6–7.

p. 246 They're having a lot of trouble On the Science Service's campaign to aid John T. Scopes's defense, its unsuccessful effort to recruit Thomas Hunt Morgan, and its difficulty in finding scientists to travel to Dayton, see Philip J. Pauly's *Biologists and the Promise of American Life: From Meriwether Lewis to Alfred Kinsey* (Princeton: Princeton University Press, 2002).

p. 247 Whatever line the defense takes On William Beebe's expedition as an effort to oppose the anti-evolution law, see Carol Grant Gould's *The Remarkable Life of William Beebe: Explorer and Naturalist* (Washington, DC: Island Press, 2004).

p. 250 I agreed with much On the anti-rationalist trend in the Industrial Workers of the World (IWW) as a reaction to evolutionary socialism's conservativism, see Mark A. Pittenger's *American Socialists and Evolutionary Thought, 1870–1920* (Madison: University of Wisconsin Press, 1993).

p. 250 I'll tell you what most On the anthropologist Robert Lowie's move from activism to strictly empirical research, see Mark A. Pittenger's "Science, Culture and the New Socialist Intellectuals before World War I," *American Studies* 28 (1987): 73–91.

p. 250 I bet Madison Grant On University of Washington anthropologist Leslie Spier's work, see Bernhard J. Stern's review "Growth of Japanese Children Born in America and Japan, by Leslie Spier," *American Journal of Sociology* 35 (1930): 675.

p. 251 Know how the Soviets For the influence of the revolution on Russian entomology, see Vladimir Vladimirovich's *Nabokov's Butterflies: Unpublished and Uncollected Writings* (Boston: Beacon Press, 2000).

p. 252 Well now, Sullivan Entomologist Karl Jordan proposed that mind-numbing intoxicants and fevers allowed one to understand a man's true motivations and subconscious better. See Jordan's "The President's Address," *Proceedings of the Royal Entomological Society of London* 5 (1931): 128–42.

p. 252 Not at all On Clarence Darrow as a materialist, see his debate with Will Durant in *Debate: Is Man a Machine? Clarence Darrow, Affirmative; Dr. Will Durant, Negative* (New York: League for Public Discussion, 1927). On La Mettrie, see Kathleen Wellman's *La Mettrie: Medicine, Philosophy, and Enlightenment* (Durham, NC: Duke University Press, 1992).

p. 253 Frank slowly smiled Frank Gray's stance on love and materialism is based on the response Anna Strunsky gave to Jack London in their jointly authored *The Kempton-Wace Letters* (New York: MacMillan, 1903).

p. 256 We both went a little mad For a sample of the view that inadequate "vocational guidance" could be dangerous for women's health, see Lewellys F. Barker's "On the Understanding and Practical Management of Nervous Patients, Particularly of the Nervous Woman," *Bulletin of the New York Academy of Medicine* 1 (1925): 404–26. On physicians like Edward Hammond Clarke who held the view that educating women the same as men would be destructive of health and fertility, see Rosalind Rosenberg's *Beyond Separate Spheres: Intellectual Roots of Modern Feminism* (New Haven, CT: Yale University Press, 1982).

p. 256 Martin did know For a description of contemporary views regarding the "generative implications of a highly intellectual life" and the eugenic obligations of women to abstain from having children if "defective," see D. Collin Wells's "Social Darwinism," *American Journal of Sociology* 12 (1907): 695–716. Also see G. Stanley Hall's *Adolescence: Its Psychology and Its Relations to Physiology, Anthropology, Sociology, Sex, Crime, Religion and Education* (New York: D. Appleton, 1915), which cited data provided by Herbert Spencer and others.

p. 257 **These fundamentalist ministers** The claim that evolution leads to Bolshevism is from Presbyterian minister Albert S. Johnson, who was quoted in the *Daily Observer* (Charlotte, NC), February 24, 1925.

p. 257 **Well, I'm not supposed** On the presumed links between atheism and immorality and how those links were eventually undermined, see Frank Turner's *Without God, Without Creed: The Origins of Unbelief in America* (Baltimore: Johns Hopkins University Press, 1985).

p. 258 **Can be a pretty good** On the sterilization of gay men in the Pacific Northwest on eugenic grounds, see Peter Boag's *Same-Sex Affairs: Constructing and Controlling Homosexuality in the Pacific Northwest* (Berkeley: University of California Press, 2003). On laws concerning, attitudes toward, and the medicalization of the newly emergent concept of "homosexuality" in the 1920s, see Marc Stein's *Rethinking the Gay and Lesbian Movement* (New York: Routledge, 2012) and section 2 of *Sexuality*, edited by Robert A. Nye (Oxford: Oxford University Press, 1999).

p. 259 **He could barely think** Darwin was quoting the German philosopher Arthur Schopenhauer. See Darwin's *The Descent of Man*, 2nd ed. (London: Murray, 1874).

Chapter 17

p. 264 **Then there was something** "Racist" was a new word in 1925, eventually replacing its predecessor "racialist" (coined in 1901).

p. 264 **Rebecca spent the night** The denial of employment to a trained nurse on grounds of her skin color (by a head nurse at St. Joseph's Hospital in Tacoma) is based on the biography of Hazel Pete. See Cary C. Collins's "A Future with a Past: Hazel Pete, Cultural Identity, and the Federal Indian Education System," *Pacific Northwest Quarterly* 92 (2000/2001): 15–28.

p. 265 **Martin couldn't explain** Will Sullivan's views of sex are based on Havelock Ellis's *Little Essays of Love and Virtue* (London: A.&C. Black, 1922).

p. 266 **Oh dear, no, that won't do** John Langdon-Davies's quip about a woman choosing the man she likes best, rather than making her decision on eugenic grounds, is from his *The New Age of Faith* (New York: Viking Press, 1925).

p. 267 **When he went inside** Trevor Kincaid's oyster work is mentioned on the same pages as the Scopes Trial in the *Courier-Journal* (Louisville, KY), July 13, 1925.

p. 268 **Then he packed a notebook** On the need for more natural history facts, and the biological station as a place for this work, see Keith R. Benson's "Experimental Ecology on the Pacific Coast: Victor Shelford and His Search for Appropriate Methods," *History and Philosophy of the Life Sciences* 14 (1992): 73–91.

p. 269 **He knew some of his colleagues** Charles Elton mentions this criticism of ecology in his textbook *Animal Ecology* (London: Sidgwick and London, 1927).

p. 270 **The Indians on this island** On the requirement that the Lummi move to the reservation in order to receive allotments in the 1880s, see Daniel L. Boxberger's "In and Out of the Labor Force: The Lummi Indians and the Development of the Commercial Salmon Fishery of North Puget Sound, 1880–1900," *Ethnohistory* 35:(1988): 161–90.

p. 272 **Later, after everyone** The work alluded to is *Contributions of Science to Religion*, edited by Shailer Mathews (New York: D. Appleton, 1924). On Mathews, see chapter 4 of Paul K. Conkin's *When All the Gods Trembled: Darwinism, Scopes, and American Intellectuals* (Lanham, MD: Roman and Littlefield, 2001)

p. 273 **Yet even Jennings** Herbert Spencer Jennings's criticisms of eugenics are from his *Prometheus: Or, Biology and the Advancement of Man* (New York: E. P. Dutton, 1925). The fact Jennings still assumed social problems had biological fixes was pointed out in reviews of *Prometheus* by both E. B. Reuter in the *American Journal of Sociology* 31 (1926): 692, and Frank Hamilton Haskins in *Social Forces* 4 (1926): 631–30. For Franz Boas's critique, see his "Eugenics," *Scientific Monthly* 3 (1916): 471–78.

p. 274 **Morgan's no trouble** On a division of labor by gender among the drosophilists, see Michael R. Dietrich and Brandi H. Tambasco's "Beyond the Boss and the Boys: Women and the Division of Labor in Drosophila Genetics in the United States, 1934–1970," *Journal of the History of Biology* 40 (2007): 509–28. On Thomas Hunt Morgan's support for female graduate students, see M. B. Ogilvie's "Inbreeding, Eugenics, and Helen Dean King (1869–1955)," *Journal of the History of Biology* 40 (2007): 467–507.

p. 275 **Oh yes, I've read my Davenport** On Herman Muller's views of eugenics and his family's struggles in Texas, see Diane Paul's "Eugenics and the Left," *Journal of the History of Ideas* 45 (1984): 567–90. Muller eventually published his views in *Out of the Night: A Biologist's View of the Future* (New York: Vanguard Press, 1935).

p. 275 **Morgan won't touch any of this** On Thomas Hunt Morgan's avoidance of public debates, including those on eugenics, see Garland Allen's *Thomas Hunt Morgan: The Man and His Science* (Princeton: Princeton University Press, 1979).

p. 275 **I suspect they regret** The stance that the welfare of the individual is more important than the welfare of the species is from Minnie Taylor's 1913 essay, discussed in Katharina Rowold's *The Educated Woman: Minds, Bodies, and Women's Higher Education in Britain, Germany, and Spain, 1865–1914* (New York: Routledge, 2011).

p. 276 **He heard the sound** In fact the radio signal was probably hard to hear on the West Coast. See James Walter Wesolowski's "Before Canon 35: WGN Broadcasts the Monkey Trial," *Journalism History* 2 (1975): 76–87.

Chapter 18

p. 278 **I believe the Bible** Josiah Gray's statement is based on Shailer Mathews's *The Faith of Modernism* (New York: Macmillan, 1924).

p. 278 **Josiah's gentle manner** The account of the trial is based on the transcripts published in *The World's Most Famous Court Trial: Tennessee Evolution Case, a Word-for-Word Report of the Famous Court Test of the Tennessee Anti-evolution Act, at Dayton, July 10 to 21, 1925, Including Speeches and Arguments of Attorneys, Testimony of Noted Scientists, and Bryan's Last Speech* (Cincinnati: National Book, 1925). For background on the trial, see Edward Larson's *Summer for the Gods: The Scopes Trial and America's Continuing Debate over Science and Religion* (New York: Basic Books, 1997), Jeffrey P. Moran's *American Genesis: The Evolution Controversies from Scopes to Creation Science* (Oxford: Oxford University Press, 2012), and Paul K. Conkin's *When All the Gods Trembled: Darwinism, Scopes, and American Intellectuals* (Lanham, MD: Roman and Littlefield, 2001). On the role of biologists in the debate, see Constance Areson Clark's *God—or Gorilla: Images of Evolution in the Jazz Age* (Baltimore: Johns Hopkins University Press, 2008).

p. 280 **The weekend** H. L. Mencken's description of fundamentalists as belonging to a different genus is from his report for the *Baltimore Evening Sun* for June 29, 1925.

p. 287 **Yes. That's true** An account of a minister's joy at converting a chief may be found in Rev. A. G. Mann's "Washington Territory. Tacoma. Thrilling Incidents," *Presbyterian Home Missionary* 15 (1886): 108.

p. 288 **I only know the stories** On the impact of World War I on Protestant missionaries' reassessment of evangelical policies on reservations, see James B. LaGrand's "The Changing 'Jesus Road': Protestants Reappraise American Indian Missions in the 1920s and 1930s," *Western Historical Quarterly* 27 (1996): 479–504.

p. 290 **She twisted her hair back** For Havelock Ellis on civilization and self-control, see his "Sexual Education and Nakedness," *American Journal of Psychology* 20 (1909): 297–317.

p. 291 **If such people** The excerpt is from George William Hunter's *A Civic Biology: Presented in Problems* (Cincinnati: American Book, 1914).

p. 297 **Madness? Plenty of biologists** Love as a disorder of mind and body, produced by passion under the stimulus of imagination, is a description provided by Jack London in his and Anna Strunsky's book *The Kempton-Wace Letters* (New York: MacMillan, 1903).

Chapter 19

p. 300 **Stuff and nonsense** On the scientists in Dayton's stance that evolution and Christianity could be reconciled and the accusation they were "devoid of moral courage," see Ira D. Cardiff's "Evolution and the Bible," *Science* 62 (1925): 111.

p. 305 **Then the defense team's ministers spoke** For an additional example of the claim that the Bible concerns moral and religious life rather than science, see R. S. Woodworth's "Similarities of Structure Show Relationship of Man and Animals: A Common Sense View of Evolution," *Science News-Letter* 7 (1925): 1–2. This particular model of the relationship between Christianity and science, which was prominent in the statements of the defense, is sometimes called a contrast or separation model.

p. 305 **Up to that point** On Kirtley Mather, see Kennard Baker Bork's *Cracking Rocks and Defending Democracy: Kirtley Fletcher Mather, Scientist, Teacher, Social Activist* (San Francisco: Pacific Division AAAS, 1994). Mather's stance can also be found in his book *Science in Search of God* (New York: Henry Holt, 1928).

Chapter 20

p. 315 **I think if we look** Martin Sullivan's stance on what was at stake in William

Jennings Bryan's criticisms, and their relation to World War I, is based on Frank Hamilton Hankins's "New Books on God, Immortality and Religious Origins," *Journal of Social Forces* 3 (1925): 747–56. For postwar criticisms of the previous century's claims regarding a progressive moral movement, see Bishop Francis J. McConnell's *Is God Limited?* (New York: Abingdon, 1924).

p. 315 **But Bryan and his fundamentalist monkeys** On publishers' removal of evolution from high school biology textbooks, including George William Hunter's *Civic Biology*, during and after the trial, see Adam Shapiro's *Trying Biology: The Scopes Trial, Textbooks, and the Antievolution Movement in American Schools* (Chicago: University of Chicago Press, 2013).

p. 316 **And I say good riddance** Ben Cardiff's complaints regarding fundamentalists are based on Robert Millikan's comments in his *Science and Life* (Pilgrim Press, 1924).

Epilogue

p. 326 **Taking in hand the bulletin description** On the influence of racist eugenics policies in the mid-1920s on previously reticent biologists' decision to speak out publicly against the movement, see Bentley Glass's "Geneticists Embattled: Their Stand against Rampant Eugenics and Racism in America during the 1920s and 1930s," *Proceedings of the American Philosophical Society* 180 (1986): 130–54. Martin Sullivan's plan to base his interrogation of race concepts in scientific taxonomy and the study of variation is based on the work of Theodosius Dobzhansky. See Paul Lawrence Farber's *Mixing Races: From Scientific Racism to Modern Evolutionary Ideas* (Baltimore: Johns Hopkins University Press, 2011).

p. 327 **Evolution may teach us** Thomas Henry Huxley's statement is from his *Evolution and Ethics: And Other Essays* (London: MacMillan, 1894). On the history of debates over defining good versus evil via an understanding of evolution, see Paul Lawrence Farber's *The Temptations of Evolutionary Ethics* (Berkeley: University of California Press, 1994).

p. 332 **Eventually Ben diagnosed** The term "parlor pink" was used to describe someone, usually either wealthy or intellectual, with leftist sympathies but who was not active in politics.

p. 333 **Then, in 1935** On the use of anti-nepotism campaigns to terminate women faculty members' contracts at the University of Washington, see Claire Paley's "Lea Miller's Protest: Married Women's Jobs at the University of Washington," and Katharine Edwards's "Married Women's Right to Work:

'Anti-Nepotism' Policies at the University of Washington in the Depression," both of which are available through the Great Depression in Washington State Project, hosted at https://depts.washington.edu/labhist/. The rule against hiring spouses remained in effect until 1971.

p. 334 **Helen looked up** For British naturalists who stopped their scientific work in order to focus on getting Jews out of Germany, and for optimism on the part of some German naturalists that Hitler's regime would reestablish the old collection networks, see Kristin Johnson's *Ordering Life: Karl Jordan and the Naturalist Tradition* (Baltimore: Johns Hopkins University Press, 2012). Charles Davenport's Eugenics Record Office was closed in 1939. Courses on eugenics continued to be taught at colleges and universities into the 1950s. It remained illegal to teach evolution in Tennessee (and several other states) until the late 1960s. For more information on subsequent trials over teaching evolution visit www.thespeciesmaker.com.